D1565489

Redox: Fundamentals, Processes and Applications

Springer
Berlin
Heidelberg
New York
Barcelona
Hong Kong
London
Milan
Paris
Singapore
Tokyo

J. Schüring · H. D. Schulz · W. R. Fischer
J. Böttcher · W. H. M. Duijnisveld (Eds.)

Redox

Fundamentals, Processes and Applications

With 110 Figures and 21 Tables

Springer

Editors:
Dr. Joachim Schüring
Prof. Dr. Horst D. Schulz
Fachbereich 5 – Geowissenschaften
Universität Bremen
Postfach 330440
D-28334 Bremen

Prof. Dr. Walter R. Fischer
Prof. Dr. Jürgen Böttcher
Institut für Bodenkunde
Universität Hannover
Herrenhäuser Strasse 2
D-30419 Hannover

Dr. Wilhelmus H.M. Duijnisveld
Bundesanstalt für Geowissenschaften und Rohstoffe
Stilleweg 2
D-30655 Hannover

ISBN 3-540-66528-5 Springer-Verlag Berlin Heidelberg New York

Library of Congress Cataloging-in-Publication Data
Redox: fundamentals, processes, and applications / J. Schüring ... [et al.], (eds.). p. cm. Includes bibliographical references and index.
ISBN 3540665285 (hardcover: alk. paper)
I. Oxidation-reduction reaction. 2. Environmental geochemistry. I. Schüring, J. (Joachim), 1962–
QE516.4.R43 2000 551.9-dc21 99-053392

Printed in Germany

Cover-Design: design & production GmbH, Heidelberg
Typesetting: Camera-ready by J. Schüring

SPIN 10673627 32/3136xz–5 4 3 2 1 0 – Printed on acid-free paper

The theory decides what is measurable.

Albert Einstein (1879-1955)

Preface

Redox – Potential or Problem?

The basic theory of redox measurements looks simple: If we have a reversible chemical reaction including electrons, the potential of an inert (platinum) electrode should reflect the actual situation of the chemical equilibrium. Therefore a high potential should indicate oxidising conditions, whereas low potentials are characteristic for reducing conditions – relative to a given redox couple.

Assuming an ideal reference electrode (potential not depending on redox conditions, low resistance, perfect contact to the water or sediment) and an ideal measuring instrument (extremely high input resistance, no voltage offset) we have to focus on how the potential of the Pt-electrode is built and what the potential-determining processes are. In detail, we have to consider some important points.

Reactions Determining the Redox Potential

If there are more than one redox reaction which contribute to the electrode charge, the total exchange current passing the electrode double layer is controlled by the activities of the oxidised and reduced species and their respective exchange currents. Therefore, the recorded potential will be a mixed potential which is mainly determined by the redox couples with the highest exchange currents.

The most important prerequisite is that the respective redox reactions have to be catalysed by the electrode material, e.g. platinum. If not (like with stable humic substances bound to the solid phase) the system does not contribute to the E_H. So what we can measure at the very best is the actual state of the dominant redox system, if it is fast enough to transfer electrons from or onto the electrode.

Many redox reactions in natural systems include a phase transformation between reduced and oxidised species. This adds another confusion to the interpretation of the measured potentials. Regarding redoximorphic soils (except those with an extremely low pH), the activity of the reduced species, e.g. Fe^{2+} or $Fe(OH)^+$, is given by the concentration and the activity coefficient, whereas the activity of the oxidised part is determined by the solubility product of the most reactive Fe(III) oxide. In this case the measured E_H reflects the electrochemical activity of dissolved Fe(II) ions.

Reversibility of the Reactions

For any E_H-calculation based on thermodynamics, the redox reactions should be reversible. In general we are not sure that this is fulfilled. Especially stereoselective enzymatic reactions are not reversible in total, whereas single steps could possibly be. Therefore we hardly can calculate E_H-values from complex redox reactions. On the other hand, besides few exceptions, it is not possible to calculate a_{ox} and a_{red} directly from measured potentials. This is the very same problem as with pH calculations: In an unknown system it is not possible to calculate the concentrations of bases or acids from measured pH.

Micro-Heterogeneity in Unsaturated Sediments and Soils

If we were going to place an electrode into bulk sediments (not soil paste or suspension) we would be confronted with another problem: we cannot exactly define the actual position of this electrode with respect to large pores or other microsite components. This way we can get realistic potentials, but the question is now: what is the sediment volume they are representative for? Unfortunately molecular oxygen predominates in many other systems, owing to the fact that it forms relatively stable Pt-oxides. Consequently the potentials of oxygen-bearing pores overcome potentials set up by other redox systems of the same sediment.

These are the main problems which we have to consider when measuring E_H-values. But, on the other hand, most of these problems occur only if we overstrain the measured values and if we try to interpret these data towards thermodynamic equilibria.

Following the NERNST-equation we can derive several empirical ranges of E_H where different redox systems are active and support bacterial metabolism. Oxygen is the electron acceptor in well aerated sediments or water. If oxygen is depleted, Mn(IV) compounds or nitrate are reduced but, due to their low concentrations, this has a negligible effect on E_H in most sediments. Fe(III) oxides are then the next components which are able to buffer E_H. After this has been reduced, sulfur and carbon redox systems are activated by strictly anaerobic metabolic processes.

This parallels strongly the interpretation of pH in sediments: Although it is not possible to calculate acid/base equilibria from pH values measured in aqueous suspension, we can distinguish soils, aquifers or marine sediments on the basis of empirical „buffer regions". So the empirical use of such figures may yield valuable information about the ecological properties of a sediment.

This book should help to find the real value of measured redox potentials of sediments and natural waters. Are they helpful when comparing them on an empirical scale, or is it possible, under certain conditions, to calculate redox equilibria, or are they only a problem and nothing else?

We wish to thank J.A.C. Broekaert, B. Hölting, U. Schwertmann, E. Usdowski and B. Wehrli who have been of great value for reviewing and commenting the chapters of this book. We are indebted to D. Maronde at the German Science Foundation (Deutsche Forschungsgemeinschaft) who has provided the financial support for a workshop on this topic within the framework of the Priority Program 546 *Geochemical Processes with long-term effects in anthropogenically affected seepage and groundwater*. Furthermore we are grateful to B. Oelkers for his competent editing of the English text.

The Editors

Contents

* * *

Chapter 7

In Situ Long-Term-Measurement of Redox Potential in Redoximorphic Soils

* * *

Chapter 8

Redox Measurements as a Qualitative Indicator of Spatial and Temporal Variability of Redox State in a Sandy Forest Soil

* * *

Chapter 9

Implementation of Redox Reactions in Groundwater Models

* * *

Chapter 10

Variance of the Redox Potential Value in Two Anoxic Groundwater Systems

* * *

Chapter 11

Redox Fronts in Aquifer Systems and Parameters Controlling their Dimensions

* * *

Chapter 12

Redox Processes Active in Denitrification

* * *

Chapter 13

Measurement of Redox Potentials at the Test Site “Insel Hengsen“

* * *

Chapter 14

Redox Reactions, Multi-Component Stability Diagrams and Isotopic Investigations in Sulfur- and Iron-Dominated Groundwater Systems

* * *

Chapter 16

Redox Zones in the Plume of a Previously Operating Gas Plant

* * *

Chapter 17

Degradation of Organic Groundwater Contaminants:
Redox Processes and E_H-Values

* * *

Chapter 18

Microbial Metabolism of Iron Species in Freshwater Lake Sediments

* * *

List of Authors

Marcus Benz
Fakultät für Biologie
Universität Konstanz
Universitätsstraße 10
D-78464 Konstanz

Jürgen Böttcher
Institut für Bodenkunde
Universität Hannover
Herrenhäuser Straße 2
D-30419 Hannover

Norbert Brandsch
Lehrstuhl für
Angewandte Geologie
Universität Karlsruhe
Kaiserstraße 12
D-76128 Karlsruhe

Andreas Dahmke
Institut für Geowissenschaften
Universität Kiel
Olshausenstraße 40-60
D-24098 Kiel

Wim H.M. Duijnisveld
Bundesanstalt für Geowissen-
schaften und Rohstoffe
Stilleweg 2
D-30655 Hannover

Markus Ebert
Institut für Geowissenschaften
Universität Kiel
Olshausenstraße 40-60
D-24098 Kiel

Matthias Eiswirth
Lehrstuhl für
Angewandte Geologie
Universität Karlsruhe
Kaiserstraße 12
D-76128 Karlsruhe

Sabine Fiedler
Institut für Bodenkunde und
Standortslehre
Universität Hohenheim
D-70593 Stuttgart

Ulrich Förstner
Arbeitsbereich
Umweltschutztechnik
TU Hamburg-Harburg
Eißendorfer Straße 40
D-21073 Hamburg

Helmuth Galster
Spessartstraße 15
D-61118 Bad Vilbel

Jörg Hencke
Fachbereich Geowissenschaften
Universität Bremen
Postfach 330440
D-28334 Bremen

Heinz Hötzl
Lehrstuhl für
Angewandte Geologie
Universität Karlsruhe
Kaiserstraße 12
D-76128 Karlsruhe

Thilo Hofmann
Institut für Geowissenschaften
Universität Mainz
Johann-Joachim-Becher-Weg 21
D-55099 Mainz

Oliver Hümmer
Institut für Wasserbau
Universität Stuttgart
Pfaffenwaldring 61
D-70550 Stuttgart

Martin Kölling
Fachbereich Geowissenschaften
Universität Bremen
Postfach 330440
D-28334 Bremen

Max Kofod
Institut für Umwelt-Geochemie
Universität Heidelberg
Im Neuenheimer Feld 234
D-69120 Heidelberg

Malte Mayer
Institut für Geowissenschaften
Universität Kiel
Olshausenstraße 40-60
D-24098 Kiel

Stefan Peiffer
Limnologische Station
Universität Bayreuth
D-95440 Bayreuth

Barbara Reichert
Geologisches Institut
Universität Bonn
Nussallee 8
D-53115 Bonn

Thomas R. Rüde
Institut für Allgemeine und
Angewandte Geologie (IAAG)
Ludwig-Maximilians-Universität
Luisenstraße 37
D-80333 München

Wolfgang Schäfer
Interdisziplinäres Zentrum für
Wissenschaftliches Rechnen
Universität Heidelberg
Im Neuenheimer Feld 338
D-69120 Heidelberg

Bernhard Schink
Fakultät für Biologie
Universität Konstanz
Universitätsstraße 10
D-78464 Konstanz

Oliver Schlicker
Institut für Geowissenschaften
Universität Kiel
Olshausenstraße 40-60
D-24098 Kiel

Mark Schlieker
Fachbereich Geowissenschaften
Universität Bremen
Postfach 330440
D-28334 Bremen

Joachim Schüring
Fachbereich Geowissenschaften
Universität Bremen
Postfach 330440
D-28334 Bremen

Ulrich Schulte-Ebbert
Institut für Wasserforschung
GmbH
Zum Kellerbach 46
D-58239 Schwerte

Horst D. Schulz
Fachbereich Geowissenschaften
Universität Bremen
Postfach 330440
D-28334 Bremen

Laura Sigg
Eidgenössische Anstalt für Wasserversorgung, Abwasserreinigung und Gewässerschutz (EAWAG)
Überlandstrasse 133
CH-8600 Dübendorf

Andreas Teichert
Institut für Bodenkunde
Universität Hannover
Herrenhäuser Straße 2
D-30419 Hannover

Jörg Thöming
Arbeitsbereich Umweltschutztechnik
TU Hamburg-Harburg
Eißendorfer Straße 40
D-21073 Hamburg

C.G.E.M. (Kees) van Beek
Kiwa N.V.
Groningenhaven 7
P.O. Box 1072
NL-3430 BB Nieuwegein

Frank von der Kammer
Arbeitsbereich Umweltschutztechnik
TU Hamburg-Harburg
Eißendorfer Straße 40
D-21073 Hamburg

Karolin Weber
Lehrstuhl für Angewandte Geologie
Universität Karlsruhe
Kaiserstraße 12
D-76128 Karlsruhe

Frank Wisotzky
Lehrstuhl für Angewandte Geologie
Universität Bochum
Universitätsstraße 150
D-44801 Bochum

Stefan Wohnlich
Institut für Allgemeine und Angewandte Geologie (IAAG)
Ludwig-Maximilians-Universität
Luisenstraße 37
D-80333 München

Chapter 1

Redox Potential Measurements in Natural Waters: Significance, Concepts and Problems

L. Sigg

1.1 Relevance of Redox Potential Measurements

The characterisation of redox conditions in natural aquatic systems is both of scientific and of practical significance. The biogeochemical processes in anoxic systems have been studied in many instances in lakes, sediments and groundwater (e.g. HERON et al., 1994; URBAN et al., 1997). To which extent the redox reactions occur in a certain system and which reduced species are present, such as reduced iron, manganese(II) and sulfide, is significant with regard to many practical problems, like use of groundwater for drinking water or remediation of contaminated sites. With regard to groundwater, the redox conditions downstream of landfills, which release large amounts of degradable compounds, need to be understood and characterised (AMIRBAHMAN et al., 1998; HERON & CHRISTENSEN, 1995).

In addition to the major redox sensitive components (O_2, $NO_3^-/N_2/NH_4^+$, SO_4^{2-}/HS^-, Mn(II)/(IV), Fe(II)/(III)), some trace elements that also undergo redox transformations behave very differently with regard to reactivity, mobility and toxicity depending on their redox state (HAMILTON-TAYLOR & DAVISON, 1995).

Redox sensitive trace elements include e.g. As(III)/(V), Se(IV)/(VI), Cr(III)/(VI). The toxic effects of these elements differ greatly for the various redox species, e.g. for Cr(III)/(VI). It is therefore of interest to predict the behaviour of these elements on the basis of the redox conditions in a certain system.

Transformation reactions of organic pollutants also strongly depend on the occurrence of oxidising or reducing conditions (SCHWARZENBACH et al., 1997; SULZBERGER et al., 1997). The reduction of nitro-aromatic compounds, for instance, may occur under iron- and sulfate-reducing conditions (SCHWARZENBACH et al., 1997).

It is therefore very important to be able to characterise the redox conditions in a given system. The electrochemical measurement of redox potential may be at first sight a very promising method, because it allows measurements in the field, *in situ* and also gives the opportunity of obtaining continuous space- and time-resolved data. The theoretical and practical difficulties associated with this method are, however, often not well-recognised, although they have already been discussed for many years (e.g. LINDBERG & RUNNELLS, 1984; STUMM & MORGAN, 1970; WHITFIELD, 1974).

In this chapter, the basic concepts and definitions concerning the redox potential, as well as the principles of its electrochemical measurement, will be briefly reviewed. The problems associated with the electrochemical redox potential measurements will be discussed and illustrated with some examples.

1.2 Thermodynamic Definition of Redox Potential

We consider here a reduction half reaction of an oxidant *Ox* to the corresponding reduced species *Red*:

$$\mathrm{Ox} + \mathrm{ne}^- = \mathrm{Red} \tag{1.1}$$

A thermodynamic constant may be defined for this reaction, using a hypothetical electron activity:

$$K = \frac{\{\mathrm{Red}\}}{\{\mathrm{Ox}\}\ \{\mathrm{e}^-\}^n} \tag{1.2}$$

where:

n = number of exchanged electrons.

To obtain a complete redox reaction, the half reaction (1.1) is combined with the oxidation of $H_{2(g)}$ to H^+, the hydrogen half reaction:

$$1/2\ H_{2(g)} = H^+ + e^- \tag{1.3}$$

for which by definition: log K = 0.
The overall redox reactions is thus:

$$Ox + \frac{n}{2} H_{2(g)} = Red + nH^+ \tag{1.4}$$

with the constant:

$$K = \frac{\{Red\}\ \{H^+\}^n}{\{Ox\}\ (p_{H_2})^{n/2}} \tag{1.5}$$

The redox potential under normal conditions is:

$$E_H^0 = \frac{2.3\ RT}{n\ F} \cdot \log K \qquad [V] \tag{1.6}$$

where:

E_H^0 = redox potential [V] (in relation to a normal hydrogen electrode) under normal conditions (all activities = 1, p_{H2} = 1 atm, $\{H^+\}$ = 1 M)
F = 1 faraday (= 96490 $C \cdot mol^{-1}$)
n = number of exchanged electrons
R = gas constant (= 8.314 $J \cdot mol^{-1} \cdot K^{-1}$)
T = temperature in K
2.3 RT/F = 0.059 V (at 25 °C).

Under other conditions the redox potential is:

$$E_H = E_H^0 + 2.3 \frac{RT}{nF} \log \frac{\{Ox\}}{\{Red\}} \tag{1.7}$$

where:

E_H = redox potential [V] (in relation to a normal hydrogen electrode).

Equation (1.7) is the NERNST equation that gives the relationship between the redox potential and the activities of the oxidised and reduced species. (For a detailed discussion see e.g. BARD & FAULKNER, 1980).

From Equation (1.2) we may also derive:

$$\log \frac{\{Red\}}{\{Ox\}} - n \log \{e^-\} = \log K \tag{1.8}$$

and

$$p\varepsilon = \frac{1}{n}\log K + \frac{1}{n}\log\frac{\{Ox\}}{\{Red\}} \tag{1.9}$$

where

$p\varepsilon$ = $-\log\{e^-\}$ defines the redox intensity.

$p\varepsilon^0$ is defined as the redox intensity for $\{Ox\}/\{Red\} = 1$:

$$p\varepsilon^0 = \frac{1}{n}\log K \tag{1.10}$$

By combining Equations (1.6) and (1.10), it is clear that:

$$E_H^0 = \frac{2.3\,RT}{F}\,p\varepsilon^0 \qquad [V] \tag{1.11a}$$

$$E_H^0 = 0.059\,p\varepsilon^0 \quad (25°C) \qquad [V] \tag{1.11b}$$

and under other conditions:

$$E_H = \frac{2.3\,RT}{F}\,p\varepsilon \qquad [V] \tag{1.12}$$

Equations (1.11) and (1.12) indicate that the redox potential E_H and $p\varepsilon$ are two equivalent scales to classify the redox reactions at equilibrium (Figure 1.1).

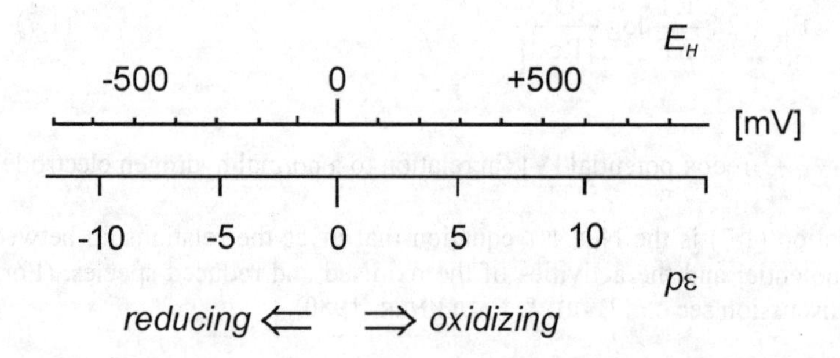

Fig. 1.1: Redox potential and pε range encountered in natural systems at near-neutral pH. Redox potentials in the positive range (up to about +800 mV) indicate the presence of strong oxidants; negative values indicate the presence of strong reductants.

It is essential to recognise that the redox potential is based on the concepts of equilibrium thermodynamics and that it can only be adequately measured at equilibrium.

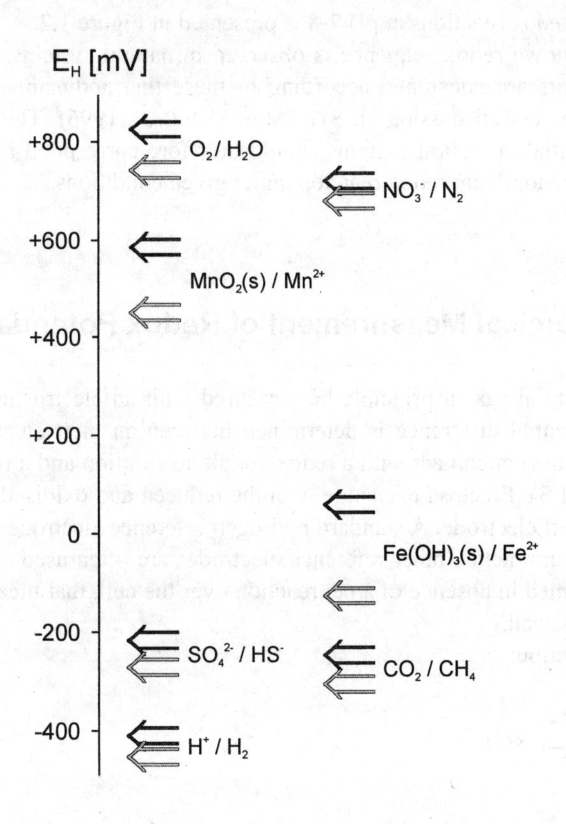

Fig. 1.2: Redox potential of some important redox reactions in natural waters at pH 7 (dark arrows) and pH 8 (light arrows). The redox potential is calculated for activities {Red} = 1 and {Ox} = 1; in the case of Mn and Fe, $\{Mn^{2+}\} = 1 \cdot 10^{-6}$ M and $\{Fe^{2+}\} = 1 \cdot 10^{-6}$ M are assumed.

1.3 Redox Potential Range in Natural Waters

The redox potential range in natural waters is limited in the negative range by the reduction of H_2O to $H_{2(g)}$ and in the positive range by the oxidation of H_2O to $O_{2(g)}$. At pH 7 to 8, the potential range reaches thus from about -400 mV to +800 mV.

The most important redox reactions in natural waters are the oxidation of organic matter and the corresponding reduction reactions: reduction of oxygen to water, nitrate to elementary nitrogen N_2, manganese(III/IV) to Mn(II), iron(III) to Fe(II), sulfate to sulfide and CO_2 to methane. The redox potential range characterising these redox reactions at pH 7-8 is presented in Figure 1.2.

The well-known redox sequence is observed in natural systems. Ideally, these various oxidants are consumed according to these thermodynamic relationships (for a more detailed discussion cf. STUMM & MORGAN, 1996). The measurement of redox potential in natural systems should therefore correspond to the potential range of the predominant redox reaction under given conditions.

1.4 Electrochemical Measurement of Redox Potential

Redox potential can in principle be measured with an electrochemical cell, in which the potential difference is determined between an inert electrode (usually made of platinum) in contact with a redox couple in solution and a reference electrode (Figure 1.3). Electron exchange with the reduced and oxidised species takes place at the inert electrode. A standard hydrogen reference electrode is represented in Figure 1.3, in practice other reference electrodes are often used. This measurement is performed in absence of a net reaction over the cell, that means at zero net current over the cell.

For any reaction:

$$\mathrm{Ox} + \mathrm{ne}^- \underset{i_{ox}}{\overset{i_{red}}{\longleftrightarrow}} \mathrm{Red} \qquad (1.13),$$

a current flows in opposite directions for the reduction and oxidation reactions. The condition of zero net current means that the rate of electrons passing over the cell in both directions (reducing and oxidising) is equal. The exchange current is the same in both directions:

$$i_{red} = i_{ox} \neq 0 \qquad (1.14)$$

where i_{red} and i_{ox} are the exchange currents of the reduction and oxidation reaction, respectively (cf. BARD & FAULKNER, 1980; STUMM & MORGAN, 1970).

However, to ensure an adequate potential measurement, the exchange current in each direction has to be different from zero, otherwise the electrode cannot detect the redox couple. This means that the exchange kinetics of reaction (1.13) have to be fast enough to produce an exchange current (BARD & FAULKNER, 1980; KEMPTON et al., 1990; PEIFFER et al., 1992). Redox potential measurements can be reliably performed in simple systems with fast exchange kinetics, like e.g. Fe(II)/(III) in acidic solution or $Fe(CN)_6^{4-}/Fe(CN)_6^{3-}$ (KEMPTON et al., 1990).

Here, however, lies a major and fundamental difficulty in measuring the redox potential in natural waters. Most of the important redox reactions mentioned above, such as O_2/H_2O, NO_3^-/N_2, NO_3^-/NH_4^+, SO_4^{2-}/HS^- and CO_2/CH_4, are very slow reactions, which occur only under the conditions of microbial catalysis. These reactions involve the exchange of several electrons and the presence of complex mechanisms. The exchange current of these reactions at an electrode is generally insufficient to establish a stable potential. Among the redox couples in natural waters, the Fe(II)/(III) couple is the most important one that exchanges electrons at an electrode at sufficient rates, especially at acidic pH (cf. BOCKRIS & REDDY, 1973; KEMPTON et al., 1990; STUMM & MORGAN, 1970).

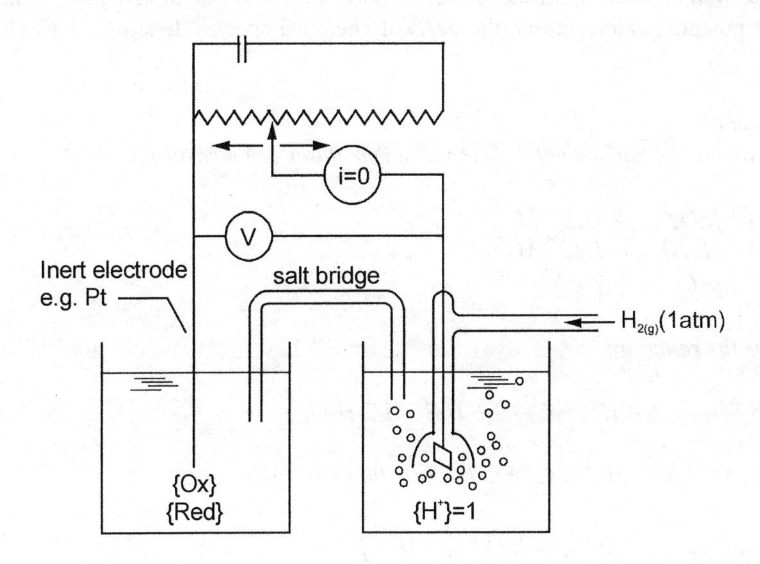

Fig. 1.3: Electrochemical cell for measurement of redox potential E_H. The potential difference is measured between an inert electrode (usually Pt) in contact with a solution containing the oxidised {Ox} and reduced {Red} species, and a reference electrode (here a standard reference hydrogen electrode, Pt in contact with $H_{2(g)}$ at 1 atm and $\{H^+\} = 1$ (at 25°C)). The potential E_H must be measured in absence of a net reaction over the cell, namely $i = 0$.

Furthermore, a reliable redox potential measurement requires that equilibrium is established not only at the electrode, but also among the various redox couples in solution. Furthermore, this condition is often not fulfilled in natural waters, because most redox reactions have slow kinetics and occur only under the influence of microbial catalysis.

The fundamental difficulties in measuring the redox potential in natural systems are thus, on the one hand, the slow kinetics of the important potential-determining redox couples at the electrodes, and on the other hand the lack of equilibrium among various redox couples.

1.5 Examples of Redox Potential Measurements

In spite of the difficulties mentioned above, redox potential measurements are often performed in natural waters. If the redox potential measurement with an electrode gives reliable results and corresponds to the equilibrium conditions in natural waters, then the measured redox potential should be in agreement with the redox potential calculated on the basis of chemical species determinations (Equation (1.7)).

Example:
In an anoxic water the following concentrations are measured:

$$[SO_4^{2-}] = 1 \cdot 10^{-5}\ M$$
$$[H_2S] = 1 \cdot 10^{-6}\ M$$
$$pH = 7$$

For the reaction:

$$1/8\ SO_4^{2-} + 5/4\ H^+ + e^- \leftrightarrow 1/8\ H_2S + 1/2\ H_2O,$$

the redox potential is defined by the following equation:

$$E_H = E_H^0 + 2.3 \frac{RT}{F} \log \frac{[SO_4^{2-}]^{1/8}\,[H^+]^{5/4}}{[H_2S]^{1/8}}$$

with $E_H^0 = 303.5\ mV$ and the concentrations mentioned above,
the following redox potential is calculated: $E_H = -206.7\ mV$

Examination of measured redox potentials shows that the measured values are oftentimes in discrepancy to the calculated redox potentials on the basis of chemical analysis.

A classical study has been published some years ago by LINDBERG & RUNNELLS, 1984, in which a large number of redox potential measurements in groundwaters has been compared to calculated values (Figure 1.4). In this study, the redox potentials have been calculated on the basis of several redox couples, of which the individual species concentrations have been determined by applying standard analytical methods. The direct electrochemical redox potential measurements have been performed in the field with Pt-electrodes. It is evident from Figure 1.4 that there is no correlation between the measured and calculated E_H values. These discrepancies indicate that the redox couples considered here do not react fast enough at electrodes to produce reliable electrochemical measurements.

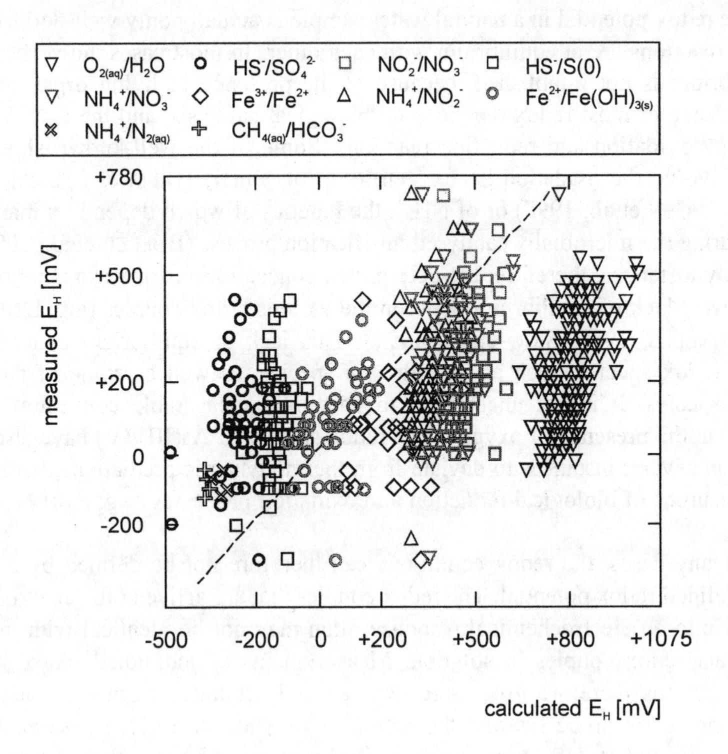

Fig. 1.4: Comparison of electrochemical field measured E_H values with calculated E_H values (calculated from measured concentrations of the redox species indicated in the legend, modified from LINDBERG AND RUNNELLS, 1984).

Some examples in the literature show, however, that more reliable electrochemical E_H-measurements may be obtained under conditions where the Fe(II)/(III) system is predominant (GRENTHE et al., 1992; MACALADY et al., 1990). In these reported cases the predominant redox reaction is:

$$Fe(OH)_3(s) + 3H^+ \leftrightarrow Fe^{2+} + 3H_2O \tag{1.15}$$

This one-electron reaction occurs at higher rates at the electrode than most of the other reactions mentioned above in Figure 1.4.

1.6 Redox Conditions in Natural Waters: Examples of Lack of Equilibrium

The redox potential in a natural water sample is actually only well-defined if all redox reactions are at equilibrium with each other. In most cases, however, redox equilibrium is not established, because of the presence of living organisms, the dependence of most redox reactions on biological catalysis, and the slow kinetics of many oxidation and reduction reactions. Some of the well-known slow reactions involve the oxidation by oxygen, e. g. of Mn(II) (DIEM & STUMM, 1984; VON LANGEN et al., 1997) or of NH_4^+, the kinetics of which depend on many factors during the microbially catalysed nitrification process (BOLLER et al., 1994). In many instances, therefore, the determined concentrations of various redox species reveal lack of equilibrium between the various redox couples (e.g. LINDBERG & RUNNELLS, 1984; SIGG et al., 1991). This lack of equilibrium concerns the major redox species, such as the nitrogen species, as well as some of the trace redox species. It is not uncommon to encounter measurable concentrations of Mn(II) in the presence of oxygen. The redox species of As(III)/(V) have also been found in several instances to deviate from thermodynamic predictions, because of the dynamics of biological reduction and oxidation processes (e.g. KUHN & SIGG, 1993).

In many cases, the redox conditions can therefore not be defined by a single well-defined redox potential. The redox couples that are active at the electrode and give rise to an electrochemical response often may not be identical with the predominant redox couples in solution. Measurements of individual redox species concentrations therefore give more detailed and accurate information about the occurring redox processes and the actual redox state in a given system. Redox species like Fe(II), HS^-, Mn(II) can easily be measured using chemical methods and probably give more reliable information about the redox state than electrochemical redox measurements bearing many uncertainties. The presence of reduced species like Fe(II) and HS^- indicates the occurrence of anoxic conditions with certainty, whereas under oxic conditions O_2 should be detectable.

1.7 Conclusions

Electrochemical measurements of redox potential in natural aquatic systems are impeded by some fundamental difficulties, namely lack of equilibrium on the one hand at the electrode and on the other hand among the various redox couples present in a given system. Some of the most important redox couples in natural waters do not react sufficiently fast at electrodes and do not contribute to stable and reliable redox potential measurements. In most cases detailed chemical analysis of natural waters with respect to several redox species therefore give a more reliable description of the redox conditions. Development of *in situ* analytical methods of reactive redox species may in the future open new perspectives in this regard.

1.8 References

AMIRBAHMAN, A.; SCHÖNENBERGER, R.; JOHNSON, C.A. & SIGG, L. (1998): Aqueous phase biogeochemistry of a calcareous aquifer system downgradient from a municipal solid waste landfill (Winterthur, Switzerland). Environ. Sci. Technol. 32: 1933-1940.

BARD, A.J. & FAULKNER, L.R. (1980): Electrochemical methods; fundamentals and applications. John Wiley & Sons, New York.

BOCKRIS, J.O.M. & REDDY, A.K.N. (1973): Modern electrochemistry. Plenum / Rosetta, New York.

BOLLER, M.; GUJER, W. & TSCHUI, M. (1994): Parameters affecting nitrifying biofilm reactors. Wat. Sci. Tech. 29: 1-11.

DIEM, D. & STUMM, W. (1984): Is dissolved Mn^{2+} being oxidized by O_2 in absence of Mn-bacteria or surface catalysts? Geochim. Cosmochim. Acta 48: 1571-1573

GRENTHE, I.; STUMM, W.; LAAKSUHARJU, M.; NILSSON, A.-C. & WIKBERG, P. (1992): Redox potentials and redox reactions in deep groundwater systems. Chem. Geol. 98: 131-150.

HAMILTON-TAYLOR, J. & DAVISON, W. (1995): Redox-driven cycling of trace elements in lakes. In: LERMAN, A.; IMBODEN, D.M. & GAT, J.R. (Eds.): Physics and chemistry of lakes. Springer-Verlag, Berlin. pp 217-263.

HERON, G. & CHRISTENSEN, T.H. (1995): Impact of sediment-bound iron on redox buffering in a landfill leachate polluted aquifer (Vejen, Denmark). Environ. Sci. Technol. 29: 187-192.

HERON, G.; CHRISTENSEN, T.H. & TJELL, J.C. (1994): Oxidation capacity of aquifer sediments. Environ. Sci. Technol. 28: 153-158.

KEMPTON, J.H.; LINDBERG, R.D. & RUNNELLS, D.D. (1990): Numerical modeling of platinum Eh measurements by using heterogeneous electron-transfer kinetics. In: MELCHIOR, D.C. & BASSETT, R.L. (Eds.): Chemical modeling of aqueous systems II. ACS, Washington DC. pp 339-349.

KUHN, A. & SIGG, L. (1993): Arsenic cycling in eutrophic Lake Greifen, Switzerland: Influence of seasonal redox processes. Limnol. Oceanogr. 38: 1052-1059.

LINDBERG, R.D. & RUNNELLS, D.D. (1984): Ground water redox reactions: an analysis of equilibrium state applied to Eh measurements and geochemical modeling. Science 225: 925-927.

MACALADY, D.L.; LANGMUIR, D.; GRUNDL, T. & ELZERMAN, A. (1990): Use of model-generated Fe^{3+} ion activities to compute Eh and ferric oxyhydroxide solubilities in anaerobic systems. In: MELCHIOR, D.C. & BASSETT, R.L. (Eds.): Chemical modeling of aqueous systems II. ACS, Washington DC. pp 350-367.

PEIFFER, S.; KLEMM, O.; PECHER, K. & HOLLERUNG, R. (1992): Redox measurements in aqueous solutions - a theoretical approach to data interpretation, based on electrode kinetics. J. Cont. Hydrol. 10: 1-18.

SCHWARZENBACH, R.P.; ANGST, W.; HOLLIGER, C.; HUG, S.J. & KLAUSEN, J. (1997): Reductive transformations of anthropogenic chemicals in natural and technical systems. Chimia 51: 908-914.

SIGG, L.; JOHNSON, C.A. & KUHN, A. (1991): Redox conditions and alkalinity generation in a seasonally anoxic lake (Lake Greifen). Mar. Chem. 36: 9-26.

STUMM, W. & MORGAN, J.J. (1970): Aquatic chemistry. Wiley-Interscience, New York.

STUMM, W. & MORGAN, J.J. (1996): Aquatic chemistry. Wiley-Interscience, New York.

SULZBERGER, B.; CANONICA, S.; EGLI, T.; GIGER, W.; KLAUSEN, J. & VON GUNTEN, U. (1997): Oxidative transformations of contaminants in natural and technical systems. Chimia 51: 900-907.

URBAN, N.R.; DINKEL, C. & WEHRLI, B. (1997): Solute transfer across the sediment surface of a eutrophic lake: I. Porewater profiles from dialysis samplers. Aquat. Sci. 59: 1-25.

VON LANGEN, P.J.; JOHNSON, K.S.; COALE, K.H. & ELROD, V.A. (1997): Oxidation kinetics of manganese (II) in seawater at nanomolar concentrations. Geochim. Cosmochim. Acta 61: 4945-4954.

WHITFIELD, M. (1974): Thermodynamic limitations on the use of the platinum electrode E_H measurements. Limnol. Oceanogr. 19: 857-865.

Chapter 2

Technique of Measurement, Electrode Processes and Electrode Treatment

H. Galster

2.1 Electron Transfers

2.1.1 Exchange Currents

The GALVANI potential of a measuring electrode is derived from the free energy ΔG of the electrode reaction. It follows from the GIBBS-HELMHOLTZ-equation:

$$-\Delta G = U \cdot n \cdot F$$

$$U = -\frac{\Delta G}{n \cdot F} \qquad (2.1)$$

where:

n = number of exchanged electrons,

F = Faraday constant.

The *Gibbs* energy ΔG can also be defined with thermodynamic data. ΔG is equal to the difference of formation heat and the product of temperature and entropy difference:

$$\Delta G = \Delta Q - T \cdot \Delta S \tag{2.2}$$

The two terms on the right side of (2.2) are made available by caloric measurements or by applying equilibrium constants. When the activities of all participants of the chemical reaction are at unity, the equilibrium constant K is defined by the *Gibbs* energy:

$$\Delta G = -RT \cdot \ln K \tag{2.3}$$

where R is the gas constant. Redox potentials, which are found in tables for many reactions, are mostly calculated from thermodynamic data (LATIMER, 1952).

The statements of these values do not imply, that they are measurable with electrochemical methods. In contrast to ion reactions, which run spontaneously, electron transfers are often inhibited. But a defined and stable redox voltage requires a reversible reaction without any inhibition.

In an apparent standstill a reversible system really behaves like a dynamic equilibrium. Due to the statistic energy distribution, forward and reverse reactions take place permanently, but they do reach a state of equilibrium in the course of time. Coexistence e.g. of Fe^{2+}- and Fe^{3+}-ions in an equilibrium (2.4) seems to be without any reaction:

$$^{*}Fe^{2+} + Fe^{3+} \leftrightarrow {}^{*}Fe^{3+} + Fe^{2+} \tag{2.4}$$

However, in molecular ranges the reaction never stands still. The so called *self-exchange* of electrons among ions of the same species can be observed by radioactive labelling as indicated by * in (2.4). The rate constant k was found to be 4 $Lmol^{-1}s^{-1}$) at 25°C (LOGAN, 1977). The life-time of an individual ion amounts to only fractions of a second.

The existence of coupled redox reactions assumes these self-exchanges, e.g. the relevant reaction in soils:

$$Fe^{3+} + Mn^{2+} \leftrightarrow Fe^{2+} + Mn^{3+} \tag{2.5}$$

The rate constant of this reaction $k_{Fe,Mn}$ can be calculated with the *MARCUS cross rule*

$$k_{Fe,Mn} = \sqrt{k_{Fe} \cdot k_{Mn} \cdot k_{Fe,Mn} \cdot f} \tag{2.6}$$

where:

k_{Fe} and k_{Mn} = self-exchange rate constants,
$k_{Fe,Mn}$ = equilibrium constant,
f = factor which is frequently equal to one.

The self-exchange is also conditional for the electron transfer between ion and electrode. In case of an equilibrium between solved ions and a Pt-electrode:

$$[Fe(H_2O)_6]^{2+} \leftrightarrow [Fe(H_2O)_6]^{3+} + e^- \text{ (metal)} \tag{2.7}$$

According to (2.7), the reaction runs in both directions. There is a negative current (i_- to the electrode) and a positive current (i_+ from the electrode). In equilibrium both currents are equal:

$$i_- = i_+ = i_0 \tag{2.8}$$

Among other factors, the amount of i_0 is a function of the electrode area. Therefore one calculates with the *exchange current density* $J_0 = i_0/cm^2$. The electrolyte concentration influences the exchange current, too. The *standard exchange current density* J_{00} has been defined on the basis of c = 1 mol/L for all reactants. In the example of Equation (2.7) it amounts $J_{00} \approx 0.4$ A/cm^2.

As long as the electrode potential is not related to the equilibrium potential, one of the two currents is greater than the other. The faster the electrode equilibrium is reached, the better the reproducibility of the redox measuring value will be.

2.1.2 MARCUS-Theory

When an electron transfer takes place between different particles in solution, it causes a change of electrical charges. This is followed by a modification of the solvent complexes. One assumption says, that an electron does not leave or enter a molecule or ion by changing its energy level (MARCUS, 1965). The energy states of the solvents must be equal in both oxidation states. In the basic state the solvent molecules build up a configuration about the central ion with a minimum of free energy. But thermal motion of the solvent molecules causes an increase of the potential energy levels.

In a harmonic vibration the potential energy E_p follows the mechanical law:

$$E_p = \frac{1}{2} m \omega^2 (x - x_0)^2 \tag{2.9}$$

where:
m = is the mass of ligands and
x = the distance from the centre.

Figure 2.1 demonstrates the oscillation of a solvent complex by a hyperbolic function of potential energy.

The total energy of a molecule or ion also contains the binding energy, the electric energy, and the kinetic energy. But this must not be considered here.

We have two hyperbolas for two oxidation states. The left one in Figure 2.1 has a smaller basic distance x_0 in the basic state. It belongs to the ion with the greater electric charge. Due to thermal oscillations two solvent complexes sometimes reach energy niveaus which are the same. This is visible as an intersection between the two hyperbolas. In this point the electron has a chance to change the complexes without crossing an energy barrier. The deeper the point of intersection, the more often electronic transfers can happen.

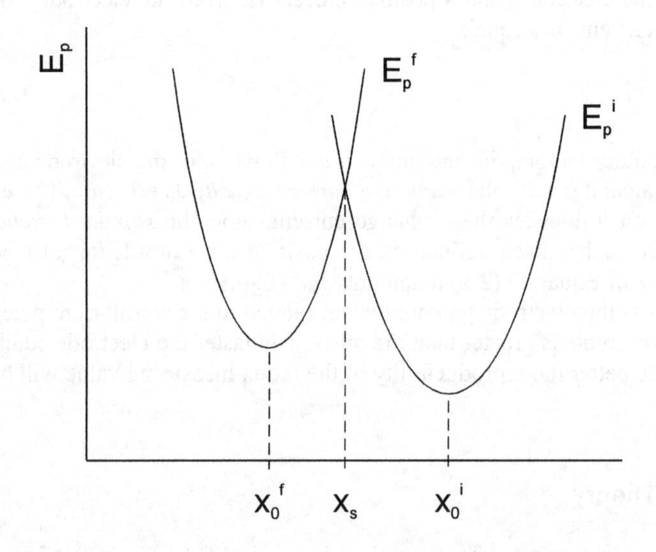

Fig. 2.1: Potential energies of a redox system with two oscillators (modified from MARCUS, 1993).

The first step in every electron transfer consists in the adsorption of two ions, or of one ion to the electrode surface, because an electron does not exist in a free state in solution. We assume for the electrode that the surface with the *HELMHOLTZ double layer* represents no barrier, since electrons do pass through by tunnelling (as a wave) without the requirement of any activation energy.

2.1.3 Practical Rules

Sometime ago experiments showed that the response time in redox measurements are smaller the less structure and solvation states must be changed by charge transfers. The exchange current is very intensive between complexed molecules or ions, if only little transformation of the structure takes place. E.g. the redox pair

$$[Fe(CN)_6]^{-4} \rightarrow [Fe(CN)_6]^{-3} + e^- \qquad (2.10)$$

has a high standard exchange current density with $J_{00} = 5$ A/cm^2. As a consequence of their large diameters the electrical field around these big ions is weak and the solvation complex possesses less structure.

One may recognise the *Principle of Least Motion* which FRANK & CONDON had derived from spectroscopic dates already in the year 1927 (see GEIB, 1941).
Some practical rules are following from theory and experience:

- Monovalent transfers are faster than multivalent transfers. The latter are only able to react stepwise.
- Electron transfers between complexated molecules or ions run faster, because their configuration remains unchanged.
- Organic redox couples are reversible, when the transfer is followed only by a negligible change of structure. No carbon bonds should be changed. Known examples are the redox couples quinone/hydroquinone and ascorbic acid/hydroascorbic acid.

Many organic redox catalysts rely on the fast electron transfer between complexated compounds.

2.2 Electrodes

Electron transfers between solution and an inert electrode result in a difference of electronic charge, which is measurable as a voltage.

2.2.1 Platinum Electrodes

In principle, every electron conducting material can be used as an electron selective electrode. In the German standard DIN 38 404-6 (1984), the redox potential is explicitly defined as the voltage which adjusts itself in an inert electronic conductor. A measuring electrode should take up the electric potential of the sample, but it should not change itself. This means that the electrode is neither allowed to react with parts of the solution nor to have any catalytic effect on an equilibrium in the solution.

Metals are inert only when their standard potential is more than 100 mV higher than the redox potential of the sample. For this reason noble metals have proven themselves as useful. Preferable are gold and platinum, see Table 2.1.

The relative exchange currents of different electrodes are estimated by current/voltage functions in various solutions. Table 2.1 demonstrates that the electrode material is important for the exchange current. Platinum is preferred to gold, although platinum has the higher exchange current density. Only with platinum does one achieve reasonable response times in diluted solutions. Other electrodes, such as silver or copper, do react with chlorides which are often present in natural waters. The fact that platinum is able to melt with glass and that it allows the manufacturing of robust electrodes in a simple way has contributed to its preference. The affinity of platinum to oxygen is less favourable, a disadvantage that must be accepted (see Chapter 2.2.2).

Tab 2.1: Comparison of gold and platinum electrodes.

Properties	*Gold*	*Platinum*
standard voltage [V]	1.42	1.2
forming oxides	no	yes
catalytic activity	rare	possible
rel. exchange current	0.3 mA/cm^2	10 mA/cm^2
melts with glass	no	yes

2.2.2 Oxygen

Molecular oxygen is less reactive at medium temperatures, because simple electron transfers are impossible without previously splitting the molecule. Metals represent catalysts which make the following reaction steps possible:

$$\begin{aligned} O_2 + 2H^+ + 2e^- &= H_2O_2 \\ H_2O_2 + H^+ + e^- &= H_2O + OH \\ OH + H^+ + e^- &= H_2O \\ \boldsymbol{O_2 + 4H^+ + 4e^-} &\boldsymbol{\leftrightarrow 2H_2O} \end{aligned} \tag{2.11}$$

Especially platinum is a catalytically active metal for oxidation with oxygen. The reason for this lies in its ability to bind oxygen in several degrees to its surface. The oxygen atoms only form a monolayer, but the binding mechanisms reach from pure adsorption to dioxides, depending on the redox potential in the surrounding solution.

HOARE (1968) has collected the standard voltages of some identified Pt-oxides, which are listed in Table 2.2. The values are valid for pH = 0. At intermediate pH-values the redox potentials are lower. The dioxides only exist under extreme pH- and redox conditions.

Tab. 2.2: Standard potentials [V] of different platinum oxides (HOARE, 1968).

Electrode	*Standard Potential*	*Electrode*	*Standard Potential*
Pt/Pt-O	0.88	$Pt/PtO_2 \cdot 4\,H_2O$	1.06
Pt/PtO	0.9	Pt/PtO_3	1.5
$Pt/Pt(OH)_2$	0.98	Pt/PtO_2	>1.6
$Pt/PtO \cdot 2\,H_2O$	1.04	Pt/Pt_3O_4	1.11
$Pt/PtO_2 \cdot 2\,H_2O$	0.96	$Pt(OH)_2/PtO_2$	1.1
$Pt/PtO_2 \cdot 3\,H_2O$	0.98	O_2	1.23

Furthermore, oxygen causes platinum to form oxides when the electrode is exposed to air during storage. An equilibrium between Pt-oxides and the solution will be adjusted in the measuring cell. If necessary, the oxide layer is diminished slowly. The reduction of Pt-oxides occurs in several steps. A possible sequence of events could be the following according to HOARE (1966):

$$
\begin{aligned}
PtO_2 + PtO(H_2O) &\rightarrow 2\,PtO(OH) \\
PtO(OH) + H^+ + e^- &\rightarrow PtO(H_2O) \\
PtO(H_2O) + 2\,H^+ + 2\,e^- &\rightarrow Pt + 2\,H_2O
\end{aligned}
\tag{2.12}
$$

The first step consists of a disproportion event which happens on the Pt-surface. The second step is a reduction by the solution. The last step requires the reaction of two electrons at the same time. In solutions which have a higher redox potential, the present Pt-oxides will be oxidised to a higher degree.

Pt-oxides are capable of conducting electrons and they do not directly influence the adjustment of redox potential, but they do increase the delay time before the equilibrium with the solution is reached. In weak redox-buffered solutions this can take a long time. So far, the higher voltage of Pt-oxides is measured.

The Pt-electrode should have a large contact surface to make extensive contact with the sample solution, but simultaneously the surface should be kept small, in microscopic dimensions, in order to minimise adsorption of oxygen. Therefore, the platinum sheet must be smooth and never platinised as is customary for hydrogen electrodes (see Chapter 2.2.4).

2.2.3
Mixed Potentials

Several reversible redox systems might react simultaneously in the same solution and contribute to an equilibrium with defined potential. When at least one of the systems is irreversible, the solution cannot arrive at an equilibrium. The galvanic potential of an electrode then has a value lying between different redox potentials. Mixed potentials can disturb redox measurements very much, because they are not reversible and mostly are not constant. They always appear, when several redox systems, which are not in equilibrium, exist simultaneously in the same solution.

Without an equilibrium every redox system has its own exchange current at the same electrode. All these currents overlay each other reciprocally (CAMMANN & GALSTER, 1996). The measurement of a redox couple which has the greater exchange current is less influenced by other systems having a smaller exchange current. An irreversible system causes only small exchange currents by itself, therefore it can influence the measurements only when it is catalysed by the electrode. Otherwise the irreversible system has no influence on the measured value.

Measurements of the redox potential of dissolved oxygen are always incorrect. This follows from the two simultaneous electrode reactions (HOARE, 1963):

$$\begin{aligned} O_2 + 4H^+ + 4e^- &= 2H_2O; \quad E_0 = 1.23\ V \\ PtO + 2H^+ + 2e^- &= Pt + H_2O; \quad E_0 = 0.88\ V \end{aligned} \tag{2.13}$$

Indeed, a value between the two values in Equation (2.13) is obtained.

Errors made in the measurement of mixed potentials can be minimised by using electrodes with the smallest catalytic effect (see Chapter 2.2.4).

2.2.4 Pre-Treatment

In the pre-treatment of Pt-electrodes it is important to distinguish between cleaning and deoxidising.

Contamination and precipitation on the surface of electrodes must be removed. In most cases hydrochloric acid is applicable as a cleaning agent. When a cleaning mixture consisting of nitric acid or chromic acid is used, an oxide layer will be produced on the surface which must be later removed. Some authors propose to glow the platinum. This procedure removes all organic contaminations, but not the metal oxides. On the contrary, metal compounds can become reduced by flame gases and metals can alloy with platinum. Such electrodes would be useless in redox measurements.

Without exception, the Pt-oxides must be removed. The redox potentials in groundwaters are smaller than that of all Pt-oxides. To accelerate the response it is necessary to clean the Pt-surface extensively from all oxides prior to each measurement. It is safest and rather easy to abrade and polish the electrode mechanically. A fine polish is useful, because a rough open surface would intensify the catalytic activity.

Some authors recommend to reduce the oxides on the electrode using chemical agents, e.g. sodium sulfite, sodium hydrogensulfite, or ascorbic acid, but every reduction turns platinum into a fine powder. Repeated applications can enlarge the catalytic activity of platinum. The same is true for cathodic reduction by applying a voltage. It is better to reduce only to the level of the probable measuring potential. Some manufactures propose the pre-treatment with chinhydrone in a buffer solution with pH = 4.01. This solution has a redox potential of E_H = 255 mV or rH = 24, and it is possible to test the measuring cell in the same time.

In the long run, one cannot avoid mechanical polishing. The polishing agent must be chemically inert and fine enough to make the Pt-surface really smooth (Corundum 1000 is recommended).

2.2.5
Design

The design of Pt-electrodes depends on their application. Electrodes in the classic design, as there are wire or sheet electrodes, tend to inflect and they can break. Today's standard is a ring which is melted onto the glass tube, its dimensions are about 6 mm in diameter and 3 mm in length. The ring is mechanically stable and can be polished without the danger of breaking. For routine measurements it is comfortable to have the reference electrode in the same rod. A combined electrode is shown in Figure 2.2A.

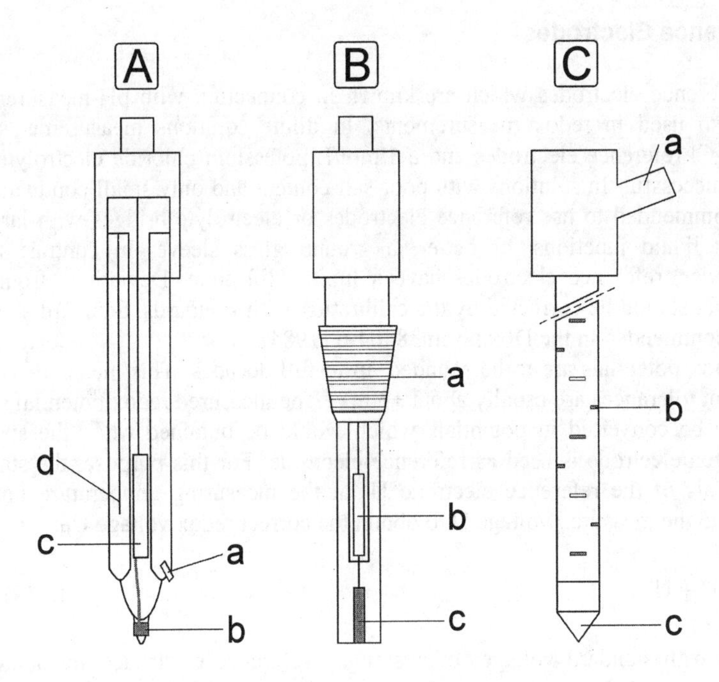

Fig. 2.2: A: Combined platinum reference electrode (a: diaphragm, b: platinum ring, c: connection plug, d: reference system). **B:** Design of a molten-in platinum rod according to KÄSS (1984) (a: conical ground joint, b: connecting plug, c: platinum rod). **C:** Design of a multi electrode according to MACHAN (1972) (a: outlet for 10-core cable, b: 10 platinum electrodes, c: puncture tip).

KÄSS (1984) has developed an exclusive design of measurements in groundwater consisting of a 2 mm diameter Pt-rod which is melted into a glass tube (Figure 2.2B). The front of the rod has an area of 3.14 mm^2 which has proven to be sufficient. Platinum and glass can be abraded and polished together for a long time.

Every sample taken to the laboratory to be measured will become changed due to the contact it makes with ambient air or the walls of the vessel. Therefore, it is important to measure the redox potential as soon as possible. Measurements in-situ are even better, if possible, one should measure in a flow-through mode. The German standard DIN 38404-6 (1984) describes a bottle with sleeve connectors for the electrodes and a thermometer. The volume should be about 500 mL. It is important, that the electrodes are tightly inserted and that no air enters.

Figure 2.2C shows a special electrode for in-situ measurements designed to be plunged into the sediments (MACHAN, 1972). Ten simple Pt-electrodes are combined to a multi-electrode. Thus, it is possible to distinguish the layers of a sediment without causing disturbances.

2.2.6 Reference Electrodes

Reference electrodes which are known in connection with pH-measurements, are also used in redox measurements. In dilute solutions measurements with Ag/AgCl reference electrodes and a 1 mol/L potassium chloride electrolyte have been successful. In solutions with poor salt content and only small conductivity it is recommended to use reference electrodes or electrolyte bridges with large diameter liquid junctions, or better, a ground glass sleeve. In contrast to Pt-electrodes, reference electrodes have a limited life-time. Deviations from their potential should be corrected by the calibration with standards. Standard solutions are recommended in the DIN norm 38404-6 (1984).

Redox potentials are to be rounded up to full decades. This means that measurement tolerances are usually about ±5 mV. The measured redox potential should always be converted in potentials which would be obtained when the standard hydrogen electrode is used as reference electrode. For this purpose, the standard potentials of the reference electrode U^0 at the measuring temperature must be added to the measured voltage U to obtain the correct redox voltage U_H:

$$U_H = U + U^0_{ret} \tag{2.14}$$

Tables with standard voltages of customary reference electrodes are obtainable from the manufacturers.

2.3 References

CAMMANN, K. & GALSTER, H. (1996): Das Arbeiten mit ionenselektiven Elektroden. Springer, Berlin, Heidelberg., New York, 26 pp.

DIN 38 404 Teil 6 (1984): Deutsche Einheitsverfahren zur Wasser- Abwasser- und Schlammuntersuchung. Physikalische und physikalisch-chemische Kenngrößen. Bestimmung der Redoxspannung. VCH, Weinheim, New York, 3 pp.

GEIB, K.H. (1941): Zum Problem Kinetik und Reaktionsmechanismus. Z. Elektrochem. 47: 761-765.

HOARE, J.P. (1963): The Normal Oxygen Potential on Bright Platinum. J. Electrochem. Soc. 110: 1019-1021.

HOARE, J.P. (1966): On the Reversible Pt Indicator Electrode. J. Electroanal. Chem. 12: 260-264.

HOARE, J.P. (1968): The Electrochemistry of Oxygen. Intersci. Publ., New York. p 19.

KÄSS, W. (1984): Redox-Messungen im Grundwasser (II). Deutsch. Gewässerkundl. Mitt. 28: 25-27.

LATIMER, W.M. (1952): The Oxidation States of the Electrode and their Potentials in Aqueous Solutions. Pentrice Hall Inc., New York p 7.

LOGAN, S.R. (1997) Grundlagen der Chemischen Kinetik. Wiley-VCH, Weinheim, New York p 108.

MACHAN, R. & OTT, J. (1972): Problems and Methods of Continous In-situ Measurements of Redox Potentials in Marine Sediments. Limnol. Oceanogr. 17: 622-626.

MARCUS, R.A. (1965): On the Theory of Electron-Transfer Reactions.VI. Unified Treatment for homogenous and Electrode Reactions. J. Chem. Phys. 43: 679-701.

MARCUS, R.A. (1993) Electron Transfer Reactions in Chemistry. Theory and Experiment. Rev. Mod. Phys. 65: 599-610.

Chapter 3

Characterisation of the Redox State of Aqueous Systems: Towards a Problem-Oriented Approach

S. Peiffer

3.1 Introduction

A great number of geochemical reactions such as the degradation of organic matter or the weathering of minerals (e.g. pyrite oxidation) are electron transfer reactions. No wonder that attempts to characterise the redox state of aqueous solutions in order to predict the occurrence of a certain redox reaction under certain geochemical boundary conditions are being made since the physical-chemical nature of these reactions has been understood. Most of these studies were performed with redox electrodes (for a review see e.g. FREVERT, 1984), which ideally consist of an inert material not involved in the redox reactions of a solution. Already back in 1946, ZOBELL was the first to use platinum electrodes to measure redox voltages at the sediment-water interface and to relate these voltages to the occurrence of certain microbial degradation processes, such as methanogenesis, regarding the redox electrode to be an easy-to-handle and a straightforward tool. This concept had been questioned in a number of articles which emphasised the kinetic constraints of redox measurements (STUMM, 1984; PEIFFER et al., 1992) and demonstrated the missing coincidence between measured and calculated redox potentials (e.g. LINDBERG & RUNNELS, 1984; see Chapter 1.5). Today, our interest in the redox state of an aqueous solution is also driven by our intention to understand reactions of environmental significance, such as the reductive or oxidative degradation of pollutants. Hitherto, no reliable method exists to unambiguously predict the fate of a certain compound in a solution with respect to the nature and the concentration of the bulk electron donors and acceptors.

In this chapter, I will therefore evaluate the limits of the use of inert redox electrodes and the interpretation of measured redox voltages in aquatic geochemistry from the perspective of electrode kinetics. A review of recent approaches to characterise the redox state of an aqueous solution will complete this chapter. The discussion is started with a conceptual comparison of the pH-value and the redox potential, a putative analogy which has caused a lot of misinterpretations.

3.2
pH and pε

Theoretically, the analogy between the pH-value and the pε-value is well-defined by the fundamentals of physical chemistry. A concise and rigorous derivation of the pε-value as a master variable for electron transfer reactions is presented in Chapter 1. The definition of the pε unambiguously describes the theoretical activity ratio of the electron acceptor and the electron donor of a redox couple (3.1)

$$p\varepsilon = p\varepsilon^0 + \lg \frac{a\,(e-Acc)}{a\,(e-Don)} \tag{3.1}$$

where:

$p\varepsilon^0$ – pε-value under standard conditions (p = 1 atm, T = 298.17 K),
a(e-Acc) = activity of the electron acceptor of a redox couple,
a(e-Don) = activity of the electron donor of a redox couple,

and refers to the law of mass action applied on the half reaction

$$Ox - e^- \Leftrightarrow Red\,.$$

$p\varepsilon^0$ corresponds to the standard potential E_H^0 of the respective half reaction according to Equation (3.2):

$$p\varepsilon^0 = \frac{E_H^0 \cdot nF}{2.303 \cdot RT} \tag{3.2}$$

where:

F = Faraday constant (96484 C/mol)
R = Gas constant (8.314 $kg \cdot m^2 \cdot s^{-2} \cdot mol^{-1} \cdot K^{-1}$)
T = absolute Temperature (K)

Thus, the pε-value denotes a theoretical electron activity $a(e^-)$,

$$p\varepsilon = -\lg a(e^-)$$

Conceptually, eq. 2.1 is identical to the equation of HENDERSON-HASSELBALCH for the pH-value

$$pH = pK_a + \lg \frac{a(H-Don)}{a(H-Acc)}, \qquad (3.3)$$

where:
pK_a = negative decadic logarithm of the acidity constant of an acid-base couple,
a(H-Acc) = activity of proton acceptor of an acid-base couple (mol/L),
a(H-Don) = activity of proton donor of an acid-base couple (mol/L).

It allows the prediction of the activity ratio of the proton acceptor and the proton donor at any acidity state of an aqueous solution.

pε-values and pH-values are used to calculate stability fields of certain redox species (e.g. GARRELS & CHRIST, 1965; NORDSTROM, 1994) and therefore predict the redox speciation form a thermodynamical point of view. The concept behind these operations is the idea to describe a chemical system if it *approaches equilibrium* at a pre-chosen pH- and pε-value, independent of the time required and the reaction pathways necessary. One aspect one can learn from these considerations is, whether a certain product will, or will not, form from a thermodynamic point of view.

There are, however, several fundamental differences between the pH-value and the pε-value. Acid-base equilibria always refer to the protolysis of the solvent, i. e. a proton acceptor reacts with water and de-protonates it, whereas a proton donor transfers protons to water to form H_3O^+. The acidity constant K_a describes to what extent an acid-base couple transfers protons to the solvent or accepts from the solvent. Hydrated protons (H_3O^+) are components of the solvent and rapidly establish equilibrium with dissolved acids or bases (this also holds true for the hydroxilated surfaces of a lot of solid phases, STUMM, 1992). Therefore, even in the absence of any dissolved acid or base, the pH is a well-defined parameter for the intensity of proton transfer reactions. Because a pH-electrode, i. e. the hydroxilated surface of a glass membrane, also establishes rapid equilibrium with water, the pH can reproducibly be measured.

In contrast, redox equilibria refer to the standard hydrogen electrode, i. e. the electron is not a component of the solvent, but of the platinum electrode, or generally spoken, the molecular orbital of a certain electron donor. Consequently, $p\varepsilon^0$ describes the tendency to accept electrons from the reactants, or donate them to the reactants. However, this does not describe the extent of electron transfer from or to the solvent. Hydrated electrons do not exist (HOSTETTLER, 1984) and therefore the pε of an aqueous solution is principally not defined in terms of a chemical entity.

This has severe implications because, contrary to a proton-sensitive sensor, an "electron"-sensitive sensor will never establish an equilibrium with water as the solvent. Therefore, measurement of pε-values (i. e. of the electron activity) in aqueous solutions are by definition not possible.

The situation changes, if water itself undergoes a redox reaction. Also in this case, no hydrated electrons exist, however, a product, such as hydrogen is formed, which is at equilibrium with water and may exist in measurable quantities. Such a situation is completely different from the pε-concept discussed above, because a certain redox species is considered in this case instead of the master variable pε.

One should have in mind at this point, that the characterisation of the redox state of an aqueous solution can, in principle, only be obtained by the measurement of the activity ratios of certain redox couples, e.g Fe^{2+}/Fe^{3+}, H_2/H_2O or H_2S/S^0. There will, however, hardly ever be a complete equilibrium between all dissolved redox species (STUMM & MORGAN, 1996) and the question arises how we should relate the information from the measurements of one individual redox couple to the bulk of the solution.

3.3 Measurement of Redox Voltages at Redox Electrodes

Redox electrodes ideally consist of an inert material, which allows an electron transfer between the oxidised and the reduced species of a redox couple through the electrode interface, without any participation of the electrode surface in the reaction. As soon as a redox electrode comes into contact with water, it will tend to equilibrate electrochemically with dissolved electroactive species. Equilibrium in this context means that the forward and backward reaction rates of the electron transfer processes via the electrode surface tend to counterbalance each other. A detailed derivation of the electrochemical principles underlying the measurements of redox voltages is given by PEIFFER et al. (1992a). The following discussion follows their theoretical treatment and the reader is referred to this article for further information.

As long as equilibrium is not established, a net current flows through the measuring circuit which is the sum of individual currents induced by the various redox couples in solution.

$$i_{net} = \sum_{j=1}^{m} i_j \qquad (3.4)$$

where:

i_{net} = total current flowing in the measuring circuit

i_j = individual current induced by a redox couple j.

The individual current is characteristic for a redox couple and depends on both, the electrode kinetics, described by the exchange current $i_{o,j}$ of a redox couple

$$i_{o,j} = k_j^0 \, C_j \, n_j \, F S$$

where:

k_j^0 = standard rate constant of a redox couple j for the electron transfer across an electrode surface (m/s). Numerical values are listed e.g. in TANAKA & TAMAMUSHI (1964), see also Table 3.1.

C_j = Concentration of the oxidised and reduced species, where $C_{ox} = C_{red}$ (mol/L).

n_j = number of electrons per molecule oxidised or reduced

S = electrode surface (m^2),

and the overvoltage η_j, i. e. the difference between the electrode voltage E and the equilibrium potential $E_{eq,j}$ corresponding to the conditions specified:

$$\eta_j = E - E_{eq,j}$$

The BUTLER-VOLMER equation combines these variables to describe the individual current:

$$i_j = i_{o,j}\left[\exp\left\{-\alpha_j \frac{n_j F}{RT}\left(E - E_{eq,j}\right)\right\} - \exp\left\{\left(1-\alpha_j\right)\frac{n_j F}{RT}\left(E - E_{eq,j}\right)\right\}\right] \quad (3.5)$$

α_j = cathodic transfer coefficient of the redox reaction j at the electrode surface (-)

The total net current i_{net} must flow in the measuring circuit and will, therefore, be restricted in magnitude by the characteristics of the amplifier in use, i. e. the input resistance of the potentiometer. The total net current can be calculated using OHM'S law:

$$i_{net} = \frac{|E|}{R} \quad (3.6)$$

with:

R: = input resistance of the potentiometer (Ω).

The input resistance of modern instrumentation will be $>10^{12}$ Ω. The value of i_{net} will therefore be $<10^{-12}$ A, if $|E| < 1$V. However, it is important to note that the total net current will never be zero.

Equations (3.4) and (3.6) can be rearranged to:

$$E = R\sum_{j=1}^{m} i_j \tag{3.7}$$

The combination of the electrochemical characteristics of the electrode-solution system (3.5) with the instrumental restrictions (3.7), allows the prediction of the electrode voltage E:

$$E = R\sum_{j=1}^{m} i_{o,j}\left[\exp\left\{-\alpha_j\frac{n_jF}{RT}\left(E-E_{eq,j}\right)\right\} - \exp\left\{\left(1-\alpha_j\right)\frac{n_jF}{RT}\left(E-E_{eq,j}\right)\right\}\right] \tag{3.8}$$

If information is available on the exchange currents $i_{o,j}$ and the transfer coefficients α_j of each redox couple in solution, Equation (3.8) can be solved by numerical methods.

From Equation (3.8) it becomes clear that the electrode voltage E is by definition a mixed potential to which all redox couples contribute, however to a variable degree. Under certain conditions the electrode voltage can be controlled by one single redox couple.

PEIFFER et al. (1992a) performed a sensitivity analysis in order to check the influence of the parameters involved in the electrode reaction on the electrode voltage. As a rule, it can be derived that a single redox couple controls the electrode voltage, if either its standard rate constant or its concentration, or both of them, are two orders of magnitude higher than that of the other redox couples.

The most important parameter seems to be the overpotential η. Even at concentrations and standard rate constants distinctly lower than those of other redox couples (less than two orders of magnitude), one single redox couple will determine the electrode voltage, if its equilibrium potential distinctly differs from that of the other ones (>0.4 V). Such a scenario, however, seems not very realistic, because redox couples with such a difference in equilibrium potential will tend to react with other dissolved or solid components[1].

In contrast, the occurrence of redox couples with high standard rate constants is not unusual at moderate concentrations (>10^{-5} mol/L). In Table 3.1 the standard rate constants of various redox couples are listed. Fe^{2+}/Fe^{3+}, Mn^{2+}/Mn^{3+}, and hydrochinone/quinone, whose rate constants are relatively high, are candidates to effectively poise the electrode voltage of a platinum electrode. Particularly in the presence of dissolved iron species, coincidence between measured redox voltages and redox potentials is reported (DOYLE, 1968; MACALADY, 1990; GRENTHE et al., 1992). From Table 3.1 one can also derive that measurement of redox voltages under oxic conditions are far from providing interpretable results because the standard rate constant of the H_2O/O_2 redox couple is much too low to compete with other redox couples e.g. organic moieties in the dissolved organic carbon, such as quinones/hydroquinones (Table 3.1). Moreover, under these conditions, the electrode material is not as inert as one may assume. VERSHININ & ROZANOV (1983) and GALSTER (see Chapter 2) discuss the various interactions between the

---

* Note, however, that for example nitrogen will never spontaneously react in the atmosphere with other substances, although the difference in free energy is sufficiently high.

electrode surface and constituents of a solution which influence the electrode voltage. Further complications are caused by the adsorption of dissolved components, which affect the electron transfer between the electrode surface and the bulk solution (BARD & FAULKNER, 1980, see also the discussion in PEIFFER et al., 1992a).

Tab 3.1: Standard rate constants of various redox couples derived from TANAKA & TAMAMUSHI (1964).

Redox couple	k^0 *[m/s]*
Fe^{2+}/Fe^{3+}	$5 \cdot 10^{-5}$
Mn^{2+}/Mn^{3+}	$1 \cdot 10^{-7}$
Hydrochinone/Quinone	$1 \cdot 10^{-6}$
H_2O/O_2	$5 \cdot 10^{-11}$

As this discussion shows, measurements of redox voltages do not provide a general parameter to *quantitatively* characterise the redox state of an aqueous solution. They *qualitatively* indicate "more reducing" and "more oxidising" conditions, because low equilibrium potentials of electrode-active redox couples will drive the redox voltages to low values and high equilibrium potentials to high values (3.8). Conclusions about the occurrence of a certain redox reaction under the conditions indicated are not allowed.

Another implication from these considerations is that measured redox voltages cannot be discussed in the framework of pε-pH diagrams derived from chemical thermodynamics. The physical nature of electrode voltages and of the redox potential is completely different. Only under certain circumstances will a redox electrode be able to detect one single redox couple and therefore fulfil the theoretical requirements derived above.

As there is typically no a-priori information existing on the composition of an aqueous solution, the use of the redox electrode will , in the case of dissolved iron, be restricted to acidic, iron-rich systems (e.g. influenced by acid mine drainage), where ferric iron also occurs at sufficiently high concentrations. In such a case, the redox electrode can be a helpful tool, because there is no sensor for dissolved iron, neither ferrous nor ferric.

3.4 New Perspectives

There is one principal lesson to be learned from the considerations discussed so far: The measurement of the concentration of one single master variable, which would allow indication of the redox state of an aqueous solution, is not possible, neither from a theoretical nor from an analytical point of view. It is even questionable whether this approach is conceptually meaningful.

Geochemical systems, where redox processes occur, are non-equilibrium systems and in most cases will never reach equilibrium* (STUMM & STUMM-ZOLLINGER 1971). As an example, organic matter should be likely to not exist in the presence of most electron acceptors (SO_4^{2-}, FeOOH, NO_3^-, etc.) found in natural systems as far as thermodynamics are concerned.

Non-equilibrium conditions imply that the nature and the extent of a redox reaction are controlled by a number of parameters, such as the gain of free energy from the redox reaction or the availability of adequate catalysts. In a lot of cases the catalytic activity is mediated by the enzymes of micro-organisms. Also, there are a number of abiotic electron transport mediators such as naturally occurring organic matter (for a review see MACALADY & TRATNYEK, 1998) or the surfaces of various minerals (e.g. STUMM, 1992).

Typical scientific questions in such a network of, mostly, coupled redox reactions refer to the extent and the rate of a certain redox reaction. Rather than measuring one single parameter it seems more meaningful to focus on the measurement of the reactants, such as sulfide and sulfate in case of sulfate reduction, and their turnover rate.

There are a lot of studies on redox processes which do well without the use of one single redox parameter. Concepts have been developed which try to answer specific questions and which do not claim to draw the complete picture of possible electron transfer reactions in an aqueous system. In the following subsections, some of these concepts will be reviewed.

3.4.1 Hydrogen Concentration as a Master Variable to Characterise Metabolic Organic Matter Degradation

One of the fundamentals in our understanding of organic-matter degradation is the concept of sequential use of electron acceptors according to the yield of free energy they provide for the metabolic activity of micro-organisms (e.g. ZEHNDER & STUMM, 1988). This concept qualitatively explains the vertical zonation of pore-water concentrations in sediments (e.g. FROEHLICH et al., 1979) or the formation of redox zones in groundwater downgrading a landfill site (GOLWER et al., 1976; BAEDECKER & BACK, 1979; LYNGKILDE & CHRISTENSEN, 1992). Inherent to this concept is that the kinetics of the organic-matter degradation rate are controlled by the energy yield of the overall reaction.

This concept, however, stands in conflict with the common experience that the terminal oxidation step is preceded by a fermentative step, which is controlled by the reactivity and the amount of the organic matter itself (e.g. BERNER, 1980; WESTRICH & BERNER, 1984; CANFIELD, 1993). During this fermentative step, small organic molecules, such as short-chain fatty acids, and H_2 are formed. The observation that these products have only very short half-lives (NOVELLI, et al.,

---

* It should be stressed at this point that maintenance of a non-equilibrium state is a pre-requisite for the existence of life.

1988) lead to the conclusion that not the oxidation step but the fermentative step is limiting for the overall degradation rate (POSTMA & JAKOBSEN, 1996). One important implication of these considerations is that the concentration of hydrogen, which is a general fermentation product (SCHLEGEL, 1992) reflects a steady-state between its formation by organic matter fermentation

$$\text{organic matter} \xrightarrow{k_{\text{fermentation}}} H_2 + \text{organic molecules}$$

and its consumption by oxidation with an adequate electron acceptor

$$H_2 + e-Acc \xrightarrow{k_{\text{oxidation}}} H_2O + e-Don$$

The steady-state concentration of H_2 then formally derives as

$$c(H_2) = \frac{k_{\text{fermentation}} \cdot c(\text{organic matter})}{k_{\text{oxidation}} \cdot c(e-Acc)}$$

Theoretical considerations by LOVLEY & GOODWIN (1988) indicate that the steady-state hydrogen concentration in sediments is controlled by the organisms catalysing the predominant terminal electron accepting process. Combining MICHAELIS-MENTEN kinetics with the growth rate of a certain type of a hydrogen utilising micro-organism (methanogenic, sulfate reducing or iron oxide reducing), they derived an expression which predicts the steady-state hydrogen concentration:

$$c(H_2) = \frac{K}{(V_{max} \cdot Y / b) - 1} \qquad (3.9)$$

where:

V_{max} = maximum H_2 uptake rate when H_2 uptake is not limited by H_2 availability (mol·(g cells)$^{-1}$),
K = H_2 concentration at which the H_2 uptake rate equals 0.5 V_{max} (mol/L),
Y = yield coefficient (g cells formed per mole H_2 consumed)
b = cell decay coefficient (time^{-1})

While V_{max} and b are regarded to be relatively constant, irrespective of the electron acceptor used in metabolism, both K and Y are strongly influenced in hydrogen metabolism by the energy yield that is available from hydrogen oxidation. While K decreases as the potential energy yield increases, Y appears to increase in hydrogen metabolism. Equation (3.9) thus predicts that the steady-state concentration of hydrogen in sediments will follow the order: methanogenic > sulfate-reducing > Fe(III)-reducing > Mn(IV)-reducing > nitrate reducing.

This concept was applied to a shallow groundwater system contaminated with organic solvents and jet fuel (CHAPELLE et al., 1996). Measurements of redox voltages and the concentrations of hydrogen were made to delineate the zonation of redox processes. Redox voltages indicated that Fe(III) reduction was the predominant redox process in the anaerobic zone and did not indicate the presence of methanogenesis or sulfate reduction. In contrast, measurements of H_2 concentrations indicated that methanogenesis predominated in heavily contaminated sediments near the water table ($H_2 \approx 7$ nmol/L) and that the methanogenic zone was surrounded by distinct sulfate-reducing ($H_2 \approx 1$ to 4 nmol/L) and Fe(III)-reducing zones ($H_2 \approx 0.1$ to 0.8 nmol/L). These results were confirmed by the distribution of dissolved oxygen, sulfate, Fe(II) and methane in the groundwater.

The conclusions that can be drawn from these considerations are as following: if one is interested in the characterisation of the metabolic pathway which is active in a certain system, the partial pressure of hydrogen seems to be a good indicator under methanogenic, sulfate-reducing or Fe(III)-reducing conditions. As was emphasised by CHAPELLE et al. (1996), H_2 concentrations in anaerobic systems should be interpreted in the context of electron-acceptor (oxygen, sulfate nitrate) availability and the presence of reduction products (Fe(II), sulfide, methane). At low $c(H_2)$ which would correspond to ferric iron reduction, great care must be taken with data interpretation because H_2 is extremely reactive and therefore the threshold value might be underestimated. Practical considerations to measure hydrogen in groundwater are made by CHAPELLE et al. (1997) and BJERG et al. (1997). Considering that the loss of hydrogen may occur during sampling and measurement in the laboratory, development of a probe for in-situ determination of hydrogen at low partial pressure seems to be necessary to further follow this concept.

3.4.2 The Partial Equilibrium Approach

The concept of fermentation to be the rate-limiting step was further refined by POSTMA & JAKOBSEN (1996). If fermentative degradation of organic matter would be the rate-limiting step, then chemical equilibrium should be approached by the terminal inorganic electron acceptors and their reduction products. In fact, fast abiotic reactions occur between various electron acceptors and reduction compounds, e. g. the reaction between ferric oxides and H_2S (PEIFFER et al. 1992b), between manganese oxide and H_2S (YAO & MILLERO, 1993) or Fe^{2+} (POSTMA, 1985). POSTMA & JAKOBSEN (1996) were able to demonstrate that the segregation of the reduction of the various electron acceptors into different redox zones is better understood as a partial equilibrium process, instead of being controlled by the overall energy yield of the various reactions. Accordingly, reduction of ferric oxides and of sulfate may proceed simultaneously over a wide range of environmental conditions. The dominant factors determining whether ferric oxide or sulfate reduction is energetically most favourable, would be the stability of the iron oxides and the pH, while the effect of the sulfate concentration is small.

This concept was confirmed by BLODAU et al. (1998) who studied the relative importance of sulfate reduction and iron (hydr)oxide reduction in the sediments of an acidic, sulfate and iron-rich, mining lake. The pore water of the sediments had a high concentration of dissolved iron, whereas the sulfide concentrations were controlled on a low level by the solubility of FeS. Although reactive iron is abundant in these sediments, the electron flow via sulfate reduction seems to exceed that of iron reduction by a factor of two. Calculation of the GIBBS free energy yield of the two processes using sediment pore-water data revealed that iron reduction was thermodynamically favoured to sulfate reduction only if amorphous iron(hydr)oxides were available. Even then the energetic advantage of iron reduction is low compared to standard conditions.

The concept of partial equilibrium does not allow to infer a certain metabolic pathway from the existence of an adequate electron acceptor. If, however, rapid abiotic redox reactions between reduction products and electron acceptors are possible (which is not the case with nitrogen compounds), then the thermodynamical equilibrium between these species (but only between these species!) seems to occur and the performance of equilibrium calculations with *measured* data would be justified.

3.4.3 The Use of the pH_2S-Value to Quantify the Redox State of Sulfidic Systems

The pH_2S-value denotes the negative logarithm of the activity of dissolved H_2S:

$$pH_2S = -\log\{a(H_2S)\}.$$

It can be measured using a combined Ag,AgI/0.2 mol/L HI/glass//Ag,Ag_2S electrode cell to which the name pH_2S-electrode cell was attributed (FREVERT & GALSTER 1978). In sulfide containing systems, equilibrium between the solution and the surface of the pH_2S-electrode cell, i.e. the Ag_2S coating of the Ag,Ag_2S half cell, is rapidly established even at very low, but buffered total sulfide activities (PEIFFER & FREVERT, 1987). In the presence of metal-sulfide precipitates with excess metal ion concentrations linearity between the pH_2S-value theoretically calculated for these solutions and the measured electromotive force was demonstrated to exist up to a pH_2S-value of 19.7 (PEIFFER & FREVERT, 1987).

It was concluded that measured pH_2S-values can be used to characterise chemical equilibrium controlled by metal-sulfide chemistry (PEIFFER, 1994). A pH_2S-scale was defined with sulfide as the strongest co-ordination partner for B-type metal ions. The co-ordination tendency of metal ions in sulfidic systems was discussed relative to the activity of H_2S. Analogously to pε-pH-diagrams, pH_2S-pH-diagrams can be established to predict metal ion speciation in sulfidic systems. A HENDERSON-HASSELBALCH-type expression was derived for the pH_2S:

$$pH_2S = pK_{SAC} + 2\,pH + \log\frac{c(MeL)}{c(L)}$$

where:

K_{SAC} = Stability constant which reflects the tendency of a sulfide acceptor, either metal ion or metal ion complex, to form a metal sulfide precipitate,

c(L),c(MeL) = concentration of metal ion or metal ion complex.

In natural systems, where iron sulfides are the most abundant metal sulfide precipitates, the pH_2S-scale is presumably limited by the solubility of amorphous, unaged FeS, where a rapid exchange equilibrium between dissolved Fe^{2+}, H_2S and the solid phase exists. The highest concentrations of Fe^{2+}, which one typically observes in anoxic and non-acidic pore waters, do not exceed 20 mmol/L. Also, the pH very seldom increases over 9 in such a system, so that these values can be used to estimate the upper limit of the pH_2S-scale. According to Equation (3.10), this limit is at a pH_2S of ≈ 11.

$$pH_2S_{u.l.} = pK_{s1,FeS} + 2\,pH - pK_{a1,H2S} + \log\{a(Fe^{2+})\} \qquad (3.10)$$

where:

$pH_2S_{u.l.}$ = upper limit of the pH_2S-scale,

$pK_{s1,FeS}$ = negative decadic logarithm of the solubility constant for the reaction: $FeS + H^+ \leftrightarrow HS^- + Fe^{2+}$. For amorphous FeS this value was determined to be 3 (DAVISON, 1991),

$pK_{a1,H2S}$ = negative decadic logarithm of the first acidity constant of H_2S (6.9, SMITH & MARTELL, 1976).

Iron sulfides of higher cristallinity, such as pyrrhotite or greigite are not at equilibrium with dissolved sulfide as long as amorphous FeS is the predominant iron sulfide phase.

The lower limit of the pH_2S-scale is given by the activity of dissolved H_2S which can hardly be higher than 100 mmol/L. Such high activity also implies weakly acidic conditions (pH ≈ 5) so that under natural conditions a pH-pH_2S-window with the co-ordinates $5 < pH < 9$ and $1 < pH_2S < 11$ exists where two important requirements concerning the intention of this paper are fulfilled:

1. The variables are thermodynamically well-defined,
2. Both variables can be rapidly and reproducibly measured.

This concept has not yet been applied to the characterisation of the redox state of a geochemical system, even so the pH_2S-electrode detects an important reduction product. Following the concept outlined by POSTMA & JACOBSEN (1996), hydrogen sulfide would be an excellent candidate to discriminate between several processes which approach partial equilibrium, such as the oxidation of organic carbon with either sulfate

$$SO_4^{2-} + 2\,CH_2O \Rightarrow 2\,HCO_3^- + H_2S$$

or ferric iron(hydr)oxide

$$4\,Fe(OH)_3 + CH_2O \Rightarrow 4\,Fe^{2+} + HCO_3^- + 7\,OH^- + 3\,H_2O\,.$$

At equilibrium, the energy yield from both processes is equal and the two reactions combine to

$$8\,Fe^{2+} + SO_4^{\;2-} + 20\,H_2O \Leftrightarrow H_2S + 8\,Fe(OH)_3 + 14\,H^+ \qquad (3.11)$$

Sulfate reducing environments are commonly found to be near equilibrium with FeS (cf. references in POSTMA & JACOBSEN, 1996) so that Equation (3.11) modifies to

$$2\,H^+ + SO_4^{\;2-} + 20\,H_2O + 8\,FeS \Leftrightarrow 9\,H_2S + 8\,Fe(OH)_3 \qquad (3.12)$$

Applying the law of mass action to Equation (3.12) one obtains:

$$9\,pH_2S + \log\{c(SO_4^{\;2-})\} - 2\,pH = \log K^* \qquad (3.13)$$

where:

K^* comprises several stability constants: the solubility of FeS and of a certain iron(hydr)oxide, the redox equilibria between sulfate and H_2S and between ferric and ferrous iron.

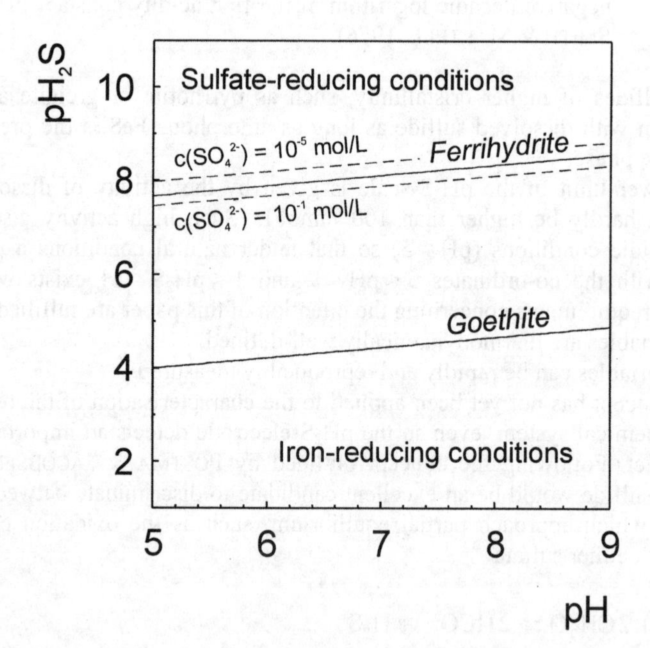

Fig. 3.1: pH_2S-pH-diagramm for the partial equilibrium between sulfate and iron reduction. Cf. text for details of the construction. Measurement of the pH_2S-value with a pH_2S electrode cell allows differentiation between iron and sulfate-reducing conditions.

Equation (3.13) was plotted in a pH_2S-pH-diagram (Figure 3.1) for goethite and for ferrihydrite. The constants used to calculate log K^* were taken from STUMM & MORGAN (1996) providing a value of 10^{-56} L^6/mol^6 for ferrihydrite and 10^{-24} L^6/mol^6 for goethite. The sulfate concentration was arbitrarily set to 10^{-3} mol/L. In case of ferrihydrite, the sulfate concentration was varied between 10^{-1} and 10^{-5} mol/L. The stability field below Equation (3.13) predicts that iron reduction occurs.

From Figure 3.1 it can be derived that sulfate reduction in the presence of ferrihydrite, an ironhydroxide of low cristallinity, is possible only at high pH_2S-concentrations, i.e. at low dissolved sulfide concentrations. Would goethite be the solid iron phase, sulfate reduction already occurs at pH_2S-values between 4 and 5. On the other hand, a pH of 7 and a pH_2S of 6 would imply that ferrihydrite is thermodynamically not stable, while goethite can still persist. The effect of the sulfate concentration on the equilibrium is of minor importance. As this example demonstrates, measurement of the pH_2S-value in anoxic systems could provide useful information about the occurrence of several redox processes. Note that pH_2S-values can be measured in-situ and could therefore complete routine measurements performed in monitoring wells or other anoxic systems.

3.4.4
The Use of Probe Compounds to Characterise the Reactivity of a Solution

A completely different approach was performed by RÜGGE et al. (1998). The biogeochemical processes controlling the reductive transformation of contaminants in an anaerobic aquifer were inferred from the relative reactivity patterns of redox-sensitive probe compounds. The concept behind this approach is based on the observation that the reductive transformation of contaminants is not necessarily linked to a predominant biogeochemical carbon mineralisation process or to one of the reduction products formed under these conditions. Rather, in-situ transformation processes are mainly controlled by the presence of catalytically active substances, such as dissolved organic matter or the surfaces of iron oxides (DUNNIVANT et al., 1992; KLAUSEN et al., 1995), whose presence cannot be derived by conventional analytical techniques.

Nitro-aromatic compounds (NAC) are regarded to be suited redox-sensitive probe compounds to characterise the type of active reductants since the relative reactivity of a given set of NACs is characteristic for the type of reductant involved in the transformation process (HADERLEIN & SCHWARZENBACH, 1995). The reactivity of substituted nitrobenzenes may vary by orders of magnitude, depending on the type and position of substituents. In their study, RÜGGE at al. (1998) used a set of five nitro-aromatic model compounds exhibiting very different reduction potentials. Their patterns of reactivity were studied in the anaerobic landfill leachate plume of a sandy aquifer and compared to the results from well-defined laboratory experiments with reductants present under sulfate-reducing and iron-reducing conditions, as were found in the plume. Despite the presence of various potential reductants (e.g. dissolved sulfide, ferrous iron, reduced organic matter, micro-organisms) the patterns of relative reactivity of the probe compounds indicated that ferrous iron associated with iron(III)(hydr)oxide surfaces was the dominant reductant throughout the anaerobic region of the plume.

Contrary to the approaches discussed in the previous sections, this approach does not intend to characterise the redox state of an aqueous system as a whole. Rather, its perspective is directed towards elucidation of fast predominant electron transfer processes with respect to certain target substances (NACs in the case of the study by RÜGGE et al. (1998)). The occurrence of these processes is certainly linked to the redox state of the system (e.g. formation of ferrous iron); the velocity with which they contribute to individual redox reactions, however, cannot be estimated from bulk parameters measured in the solution or in the solid phase.

3.5 Summary and Conclusion

The perspectives discussed in this chapter reflect different approaches which are valid only under certain circumstances: hydrogen seems to be suited to indicate the predominant metabolic pathway of organic carbon degradation which is typically called *primary redox reaction* (WANG & VAN CAPELLEN, 1996). Its application seems to be limited to the discrimination of methanogenesis, iron-reducing, and sulfate-reducing conditions. It needs to be studied whether an improved hydrogen sensor technique will allow detection of manganese- or nitrate-reducing conditions. The concentration of hydrogen will not provide information about secondary redox reactions such as the reaction of H_2S with ferric oxides.

The concept of partial equilibrium refers only to rapid secondary redox reactions between the reactants of primary redox reactions and other redox-active compounds. The advantage here is that measured concentrations of reactants can be used to calculate chemical equilibria. Future studies should be directed to elucidate partial equilibria and strengthen this concept. The use of the pH_2S-value as a master variable in sulfidic systems to describe partial equilibria may be a helpful tool and should be critically evaluated.

Application of probe compounds seems to be a promising approach to identify redox reactions between reactants which are neither at partial equilibrium nor directly linked to one of the primary redox reactions. For example, the formation of ferrous iron cannot be attributed to one single primary or secondary redox reaction. Depending on its coordinative environment, ferrous iron can be a very powerful reductant, being involved in a lot of redox reactions. Probe compounds may help to elucidate its reactivity as discussed in the example above. Similar approaches to characterise the reactivity of ferric (hydr)oxides or organic matter are urgently needed.

The main conclusion from these considerations is that there is no theoretical basis for a general parameter which would allow the characterisation of the redox state of aqueous systems. Consequently, a standard method is missing. The study of redox reactions can only be approached by a careful, problem-oriented analysis of the scientific question. More strategies similar to those discussed in this chapter that will increase our understanding of individual redox processes, have to be developed in the future. In-situ measurement techniques for redox active compounds such as the pH_2S electrode cell (e.g. for Fe^{2+} or Mn^{2+}) would be a valuable support for such studies.

3.6 References

BAEDECKER, M.J. & BACK, W. (1979): Modern marine sediments as a natural analog to the chemically stressed environment of a landfill. J. Hydrol. 43: 393-414.

BARD, J.A. & FAULKNER, L.R. (1980): Electrochemical Methods - Fundamentals and Applications. Wiley, New York.

BERNER, R.A. (1980): Early Diagenesis - a Theoretical Approach. Princeton University Press, Princeton.

BJERG, P.L.; JACOBSEN, R.; BAY, H.; RASMUSSEN, M.; ALBRECHTSEN, H.J. & CHRISTENSEN, T.H. (1997): Effects of sampling well construction on H_2 measurements made for characterisation of redox conditions in a contaminated aquifer. Environ. Sci. Technol. 31: 3029-3031.

BLODAU, C.; HOFFMANN, S.; PEINE, A. & PEIFFER, S. (1998): Iron and sulfate reduction in the sediments of acidic mine lake 116 (Brandenburg, Germany): Rates and geochemical evaluation. Water Air Soil Poll., in press.

CANFIELD, D.E. (1993): Organic matter oxidation in marine sediments. In: WOLLAST, R.; MACKENZIE, F.T. & CHOU, L. (Eds.) Interactions of C, N, P and S Biogeochemical Cycles and global change. NATO ASI Series, Springer, pp 333-363.

CHAPELLE, F.H.; HAACK, S.K.; ADRIAENS, P.; HENRY, M.A. & BRADLEY, P.M. (1996): Comparison of Eh and H_2 measurements for delineating redox processes in a contaminated aquifer. Environ. Sci. Technol. 30: 3565-3569.

CHAPELLE, F.H.; VROBLESKY, D.A.; WOODWARD, J.C. & LOVLEY, D.R. (1997): Practical considerations for measuring hydrogen cocentrations in freshwaters. Environ. Sci. Technol. 31: 2873-2877.

DAVISON, W. (1991): The solubility of iron sulphides in synthetic and natural waters at ambient temperature. Aquatic Sciences 53: 309-329.

DOYLE, R.W. (1968): The origin of the ferrous ion-ferric oxide Nernst potential in environments containing dissolved ferrous iron. Am. J. Sci. 266: 840-859.

DUNNIVANT, F.M.; SCHWARZENBACH, R.P. & MACALADY, D.L. (1992): Reduction of substituted nitrobenzenes in aqueous solutions containing natural organic matter. Environ. Sci. Technol. 26: 2133-2141.

FREVERT, T. (1984): Can the redox conditions in natural waters be predicted by a single parameter? Schweiz. Z. Hydrol. 46: 269-290.

FREVERT, T. & GALSTER, H. (1978): Schnelle und einfache Methode zur In-situ Bestimmung von Schwefelwasserstoff in Gewässern und Sedimenten. Schweiz. Z. Hydrol. 40: 199-208.

FROELICH, P.N.; KLINKHAMMER, G.P.; BENDER, M.L.; LUEDTKE, N.A.; HEATH, G.R.; CULLEN, D. & DAUPHIN, P. (1979): Early oxidation of organic matter in pelagic sediments of the eastern equatorial Atlantic: suboxic diagenesis. Geochim. Cosmochim .Acta 43: 1075-1090.

GARRELS, R.M. & CHRIST, C.L. (1965): Solutions, Minerals and Equilibria. Freeman, San Francisco.

GOLWER, A.; KNOLL, K.H.; MATTHEß, G.; SCHNEIDER, W. & WALLHÄUSER, K.H. (1976): Belastung und Verunreinigung des Grundwassers durch feste Abfallstoffe. Abhandlungen des hessischen Landesamtes für Bodenforschung 73.

GRENTHE, J.; STUMM, W.; LAAKSUHARJU, M.; NILSSON, A.C. & WIKBERG, P. (1992): Redox potentials and redox reactions in deep groundwater systems. Chem. Geol. 98: 131-150.

HADERLEIN, S.B. & SCHWARZENBACH, R.P. (1995): Environmental processes influencing the rate of abiotic reduction of nitroaromatic compounds in the subsurface. In: SPAIN, J.C. (Ed.) Biodegradation of Nitroaromatic Compounds. Plenum Press, New York, pp 199-225.

HOSTETTLER, J.D. (1984): Electrode electrons, aqeuous electrons, and redox potentials in natural waters. Amer. J. Sci. 284: 734-759.

KLAUSEN, J.; TRÖBER, S.P.; HADERLEIN, S.B. & SCHWARZENBACH, R.P. (1995): Reduction of substituted nitobenzenes by (Fe(II) in aqueous mineral suspensions. Environ. Sci. Technol. 29: 2396-2404.

LINDBERG, R.D. & RUNNELS, N.N. (1984): Groundwater redox reactions: an analysis of equilibrium state spplied to E_h measurements and geochemical modeling. Science 225: 925-927.

LOVLEY, D.R. & GOODWIN, S. (1988): Hydrogen concentrations as an indicator of the predominant terminal electron accepting reactions in aquatic sediments. Geochim. Cosmochim. Acta 52: 2993-3003.

LYNGKILDE, J. & CHRISTENSEN, T.H. (1992): Redox zones of a landfill leachate pollution plume (Vejen, Denmark). J. Cont. Hydrol. 10: 273-289.

MACALADY, D.L. & TRATNYEK, P.G. (1998): Oxidation/reduction reactions in aquatic systems. In: BRUSSEAU, B. & MACKAY, D. (Eds.), in press.

MACALADY, D.L.; LANGMUIR, D.; GRUNDL, T. & ELZERMAN, A. (1990): Use of model-generated Fe^{3+} ion activities to compute E_h and ferric oxyhydroxide solubilities in anaerobic systems. In: MELCHIOR, D.C. & BASSETT, R.L. (Eds.) Chemical Modeling of aqueous Systems II, ACS Symposium Series. Vol 416, American Chemical Society, Washington, pp 350-367.

NORDSTROM, D.K. & MUNOZ, J.L. (1994): Geochemical Thermodynamics. Blackwell , Boston.

NOVELLI, P.C.; MICHELSON, A.R.; SCRANTON, M.I.; BANTA, G.T.; HOBBIE, J.E. & HOWARTH, R.W. (1988): Hydrogen and acetate in two sulfate-reducing sediments: Buzzards Bay and Town Cove, Mass. Geochim. Cosmochim. Acta 52: 2477-2486.

PEIFFER, S. (1994): Predicting trace metal speciation in sulphidic leachates from anaerobically digested waste material by use of the pH_2S value as a mastervariable. J. Cont. Hydrol. 16: 289-313.

PEIFFER, S. & FREVERT, T. (1987): Potentiometric determination of heavy metal sulphide solubilities using a pH_2S (glass/Ag^0,Ag_2S) electrode cell. Analyst 112: 951-954.

PEIFFER, S.; KLEMM, O.; PECHER, K. & HOLLERUNG, R. (1992a): Redox measurements in aqueous solutions - a theoretical approach to data interpretation, based on electrode kinetics. J. Cont. Hydrol. 10: 1-18.

PEIFFER, S.; DOS SANTOS AFONSO, M.; WEHRLI, B. & GÄCHTER, R. (1992): Kinetics and Mechanism of the Reaction of H_2S with Lepidocrocite. Environ. Sci. Technol. 26: 2408-2413.

POSTMA, D. (1985): Concentration of Mn and separation from Fe in sediments. I. Kinetics and stoichiometry of the reaction between birnessite and dissolved Fe(II) at 10°C. Geochim. Cosmochim. Acta 49: 1023-1033.

POSTMA, D. & JAKOBSEN, R. (1996): Redox zonations: Equilibrium constraints on the Fe(III)/SO-reduction interface. Geochim. Cosmochim. Acta 60: 3169-3176.

RÜGGE, K.; HOFSTETTER, T.B.; HADERLEIN, S.B.; BJERG, P.L.; KNUDSEN, S.; ZRAUNIG, C.; MOSBAEK, H. & CHRISTENSEN, T.H. (1998): Characterization of predominant reductants in an anaerobic leachate-contaminated aquifer by nitoaromatic probe compounds. Environ. Sci. Technol. 32: 23-31.

SCHLEGEL, H.G. (1992): Allgemeine Mikrobiologie, Thieme, Stuttgart.

SMITH, R.M. & MARTELL, A.E. (1976): Critical Stability Constants, Inorganic Complexes. Vol 4, Plenum, New York.

STUMM, W. (1984): Interpretation and measurement of redox intensity in natural waters. Schweiz. Z. Hydrol. 46: 291-296.

STUMM, W. (1992): Chemistry of the Solid Water Interface. Wiley, New York.

STUMM, W. & MORGAN, J.J. (1996): Aquatic Chemistry. Wiley, New York.

STUMM, W. & STUMM-ZOLLINGER, E. (1971): Chemostatis and Homeostasis in Aquatic Ecosystems; Principles of Water Pollution Control. In: HEM, J.D. (Ed.) Nonequilibrium Systems in Natural Water Chemistry. Vol 106, ACS Advances in Chemistry Series, Washington, pp 1-29.

TANAKA, N. & TAMAMUSHI, R. (1964): Kinetic parameters of electrode reactions. Electrochim. acta 9: 936-989.

VERSHININ, A.V. & ROZANOV, A.G. (1983): The platinum electrode as an indicator of redox environment in marine sediment. Mar. Chem. 14: 1-15.

WANG, Y. & VAN CAPPELLEN, P. (1996): A multicomponent reactive transport model of early diagenesis: Application to redox cycling in coastal marine sediments. Geochim. Cosmochim. Acta 60: 2993-3014.

WESTRICH, J.T. & BERNER, R.A. (1984): The role of sedimentary organic matter in bacterial sulfate reduction: The G model tested. Limnol. Oceanogr. 29: 236-249.

YAO, W. & MILLERO, F.J. (1993): The rate of sulfide oxidation by δ-MnO_2 in seawater. Geochim. Cosmochim. Acta 57: 3359-3365.

ZEHNDER, A.J.B. & STUMM, W. (1988): Geochemistry and biogeochemistry of anaerobic habitats. In: ZEHNDER, A.J.B. (Ed.) Biology of Anaerobic Microorganisms. Wiley, New York, pp 1-38.

ZOBELL, C.E. (1946): Studies on redox potentials of marine sediments. Bull. Am. Assoc. Pet. Geol. 30: 477-513.

Chapter 4

Comparison of Different Methods for Redox Potential Determination in Natural Waters

M. Kölling

4.1 Introduction

In geochemical equilibrium model programs such as PHREEQE (PARKHURST et al., 1990) PHREEQC (PARKHURST, 1995) the redox potential is a major variable strictly affecting the species distribution and mineral stability of both, redox-sensitive dissolved species and solid phases. It is therefore necessary to assess the accuracy of pε determinations by redox probes. In addition, there are often inconsistencies to be found in calculated pε values. In different geochemical computer model programs, the pε may be calculated from the concentrations of various redox couples. These values often vary by hundreds of millivolts in one single water analysis. Therefore, the reliability of pε calculations is discussed.

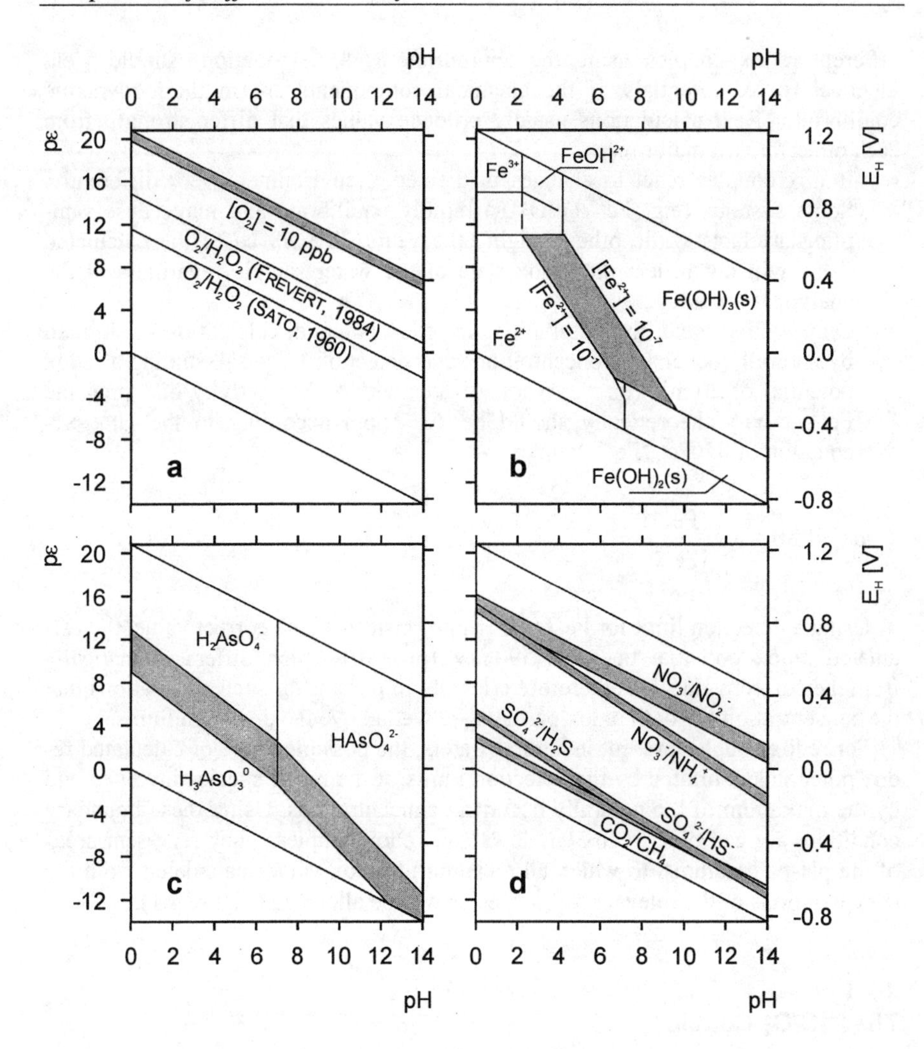

Fig. 4.1: "Redox windows" (shaded areas) for different redox couples:
(a) Oxygen = 10 ppb to saturation; **(b)** $[Fe^{2+}] = 10^{-1}$ to 10^{-7} mol/L, $[Fe^{3+}]/[Fe^{2+}] = 10^{-2}$ to 10^{2}; **(c)** $[As(V)]/[As(III)] = 10^{4}$ to 10^{-4}; **(d)** $[NO_3^-]/[NO_2^-] = 10^{4}$ to 10^{1}, $[NO_3^-]/[NH_4^+] = 10^{-3}$ to 10^{3}, $[CO_2]/[CH_4] = 10^{-3}$ to 10^{3}, $[SO_4^{2-}]/[H_2S]$ 10^{3} to 10^{-3} (modified from KÖLLING, 1986).

4.2 Redox Potential Calculations

In natural waters there are numerous ions that participate in different redox reactions. Theoretically, the redox potential calculated from the activity ratio of different redox couples using the appropriate NERNST-equations should yield identical redox potentials, if the chemical components are in thermodynamic equilibrium. Real calculations usually produce values that differ strongly from each other for two major reasons:

- Redox couples react to changes in the redox environment very differently. Some systems (e.g. $Fe^{2+}/Fe(OH)_3$) rapidly equilibrate and may cause sampling artefacts while others might take years. In both cases the calculated value will not reflect the redox state of the water sample regardless of the analytical accuracy.
- On the other hand, the possibilities in redox potential calculation are limited by naturally occurring concentrations and detection limits. Assuming a redox potential of 20 mV (pε = 3.4) for a water with a Fe^{2+} activity of 1 ppm the Fe^{3+}-activity theoretically should be $10^{-9.6}$ ppm according to the NERNST-equation of the Fe^{2+}/Fe^{3+}-couple:

$$p\varepsilon = 13.0 + \log\frac{[Fe^{3+}]}{[Fe^{2+}]} \quad (4.1)$$

Using a detection limit for Fe^{3+} (e.g. 1 ppb) instead of the correct value, the calculated redox potential becomes 590 mV (pε = 10) which differs dramatically from the correct value. It is therefore critically important for such calculations that the concentrations of both redox partners are well above the detection limits.

For redox couples present in natural waters, the possible range of calculated redox potentials is limited by the detection limits of the redox species involved and by the maximum of the naturally occurring concentrations. Using these boundary conditions we can plot "redox windows" for redox couples. They represent areas of the pH-pε diagrams, to which all meaningful redox values calculated from the concentrations of the relevant redox species will be allocated (Figure 4.1).

4.2.1 The H_2O/O_2-Couple

The standard potential of the H_2O/O_2-couple is very high (E_H^0 = 1230 mV, $p\varepsilon^0$ = 20.78).

$$p\varepsilon = 20.78 - pH + \frac{1}{4}\log pO_2 \quad (4.2)$$

Assuming a detectable amount of oxygen (i.e. 10 ppb) the calculated redox potential at pH = 7 becomes 760 V (pε = 12.9). Since such high redox values are usually not observed, SATO (1960) stated that the oxygen reduction in water occurs via a metastable H_2O_2 phase which controls the redox potential of the solution.

The NERNST-equation for this couple is:

$$p\varepsilon = 11.7 - pH + \frac{1}{2}\log\frac{pO_2}{[H_2O_2]} \tag{4.3}$$

The activity of the metastable H_2O_2 is not detected in routine analyses but the activity ratio may not become smaller than unity since H_2O_2 is formed from O_2, hence

$$p\varepsilon \geq 11.7 - pH \tag{4.4}$$

should be the minimum redox potential of water containing oxygen (Figure 4.1a).

According to FREVERT (1984) measurements in well aerated waters show pε values which may be explained with a predominance of the O_2/H_2O_2-couple assuming a H_2O_2 activity of 10^{-7} M estimated on the basis of available data. On the basis of this estimate the measured pε values should follow

$$p\varepsilon = 14.9pH + \frac{1}{2}\log pO_2 \qquad [H_2O_2] = 10^{-7}\ M \tag{4.5}$$

The calculation of redox potentials from the oxygen concentration "according to SATO (1960)" is erroneous in some geochemical equilibrium model programs inasmuch as 11.7 is used instead of 20.78 as the $p\varepsilon^0$ for the H_2O/O_2-couple, a value which has never been proposed by SATO (1960).

4.2.2
Fe-Species

In Figure 4.1b the redox windows of some Fe-redox couples are shown. In natural waters with normal pH the pε may be calculated from:

$$p\varepsilon = 16.5 - 3pH - \log[Fe^{2+}] \tag{4.6}$$

4.2.3
As-Species

CHERRY et al. (1979) introduced the calculation of redox potentials from the concentrations of As (III) and As(V). Using an appropriate method as described by SHAIK & TALLMANN (1978) species-specific analyses of As(III) and As(V) down to ng-concentration levels allow reliable calculations of redox potentials. Arsenic responds to changes in redox conditions within days or weeks. This allows sampling and preservation of redox states but also assures sufficiently quick adaptation to changes of redox conditions. The As-"redox window" is in a region where measured values are often found (Figure 4.1c).

$$p\varepsilon = 11 - \frac{3}{2}pH + \frac{1}{2}\log\frac{[H_2AsO_4^-]}{[H_3AsO_3]} \qquad \text{for pH} < 7$$

$$p\varepsilon = 14.5 - 2\,pH + \frac{1}{2}\log\frac{[HAsO_4^{2-}]}{[H_3AsO_3]} \qquad \text{for pH} > 7 \tag{4.7}$$

4.2.4
S-Species

Under reducing conditions sulfate is reduced to sulfide. The redox potential may be calculated with the NERNST-equations:

$$p\varepsilon = 5.12 - \frac{5}{4}pH + \frac{1}{8}\log\frac{[SO_4^{2-}]}{[H_2S]} \qquad \text{for pH} < 7$$

$$p\varepsilon = 4.25 - \frac{9}{8}pH + \frac{1}{8}\log\frac{[SO_4^{2-}]}{[HS^-]} \qquad \text{for pH} > 7 \tag{4.8}$$

The sulfate/sulfide-couple is often far from equilibrium. From the NERNST-equations it may be seen that, due to eight electron transfers, the changes in the calculated redox potential only amount to 1/8 pε unit or 7 mV per order of magnitude change in the sulfate to sulfide ratio. As a consequence, the sulfur "redox window" (Figure 4.1d) is very narrow (pε = 3.2 to 4 or E_H = 190 to 235 mV at pH = 7).

4.2.5
N-Species

Several redox reactions may be formulated for nitrogen species. In natural waters the nitrate/nitrite-couple and the nitrate/ammonia-couple are often used:

$$p\varepsilon = 14.15 - pH + \frac{1}{2}\log\frac{[NO_3^-]}{[NO_2^-]}$$

$$p\varepsilon = 14.9 - \frac{5}{4}pH + \frac{1}{8}\log\frac{[NO_3^-]}{[NH_4^+]} \tag{4.9}$$

4.2.6
The CO_2/CH_4-Couple

In strong reducing environments CO_2 is reduced to CH_4 and the redox potential may be calculated with the following equation:

$$p\varepsilon = 2.87 - pH + \frac{1}{8}\log\frac{[CO_2]}{[CH_4]} \tag{4.10}$$

4.3
Redox Potential Measurements

The redox potential is measured by using redox probes. The potential difference is determined between a metal electrode and a reference electrode whereby both are immersed in the solution to be measured. The redox potential is generated by transfer of electrons from the metal electrode into the solution and vice versa. The net current from this process depends on the redox species, its concentration in solution and the material of the electrode. The redox potential equilibrates where the absolutes of the cathodic and anodic currents are equal and the net current becomes zero. Net current graphs drawn for different redox couples should theoretically meet at one redox potential. (Figure 4.2a). In natural waters, redox couples are usually far from equilibrium. In this case electrodes show mixed potentials. The mixed potential is determined by redox couples with the steepest net current curve (Figure 4.2b). Redox couples with a smaller slope in this curve produce a net current is close to zero over a wide redox potential range, which results in a poor stability as far as the measurements are concerned.

Natural waters are usually very dilute solutions in terms of the concentration of redox-sensitive species. Redox probes only respond to processes which quickly and reversibly occur at the metal electrode surface. According to STUMM &

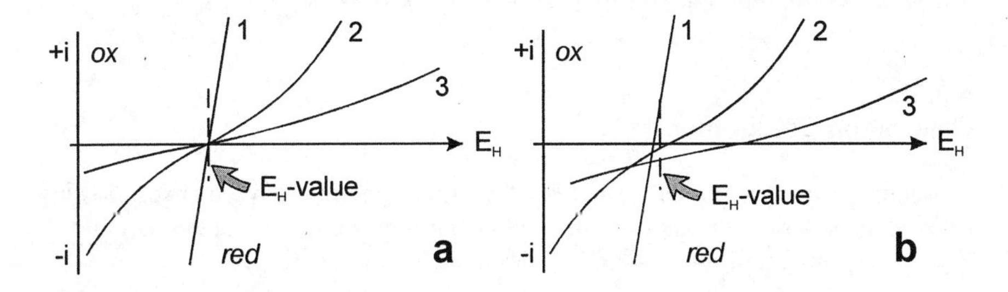

Fig. 4.2: Net current graphs for a set of three redox couples:
(a) equilibrium between redox couples; **(b)** natural situation of non-equilibrium between redox couples. The redox probe will show the potential, where the sum of all three net current curves becomes zero (dashed lines, modified from KÖLLING, 1986).

MORGAN (1996), this condition does not hold for nitrate, nitrite, sulfide, ammonia and the CO_2/CH_4-system.

FREVERT (1984) clearly distinguishes between the log of the electron activities calculated from electrode measurements which he refers to as pe, and the pε which is conceptually defined as the redox intensity. The pe and pε values are equal when equilibria exist between different redox couples (Figure 4.2a) and between the solution and the electrode. A detailed theoretical background on the interactions between electrode, aqueous electron activities, and redox potentials is given by HOSTETTLER (1984).

4.3.1 Metal Electrodes

Redox probes often use platinum as the metal electrode. In solutions with high oxygen concentrations chemi-sorption of oxygen to the platinum surface results in a redox reaction Pt/Pt-O which is induced by the electrode itself and affects the measured values (WHITFIELD, 1974)

4.3.2 Reference Electrodes

The reference to which all potentials are related is the standard hydrogen electrode (SHE). Today, a Ag/AgCl electrode is used in most laboratory and field measurements. The Ag/AgCl reference system has its own potential related to the SHE. This potential has to be added to all measured values

$$\begin{aligned} &E_H(SHE) = E_{H(measured)} + E_{H(reference)} \\ &Ag/AgCl: E_{H(reference)}\,[mV] = 207 + 0.7(25 - t) \end{aligned} \tag{4.11}$$

where t is the temperature in [°C].

Most combined redox probes do not have a durable label showing the reference system or the appropriate electrolyte solution to be used.

4.3.3 Calibration Solutions

Redox probes are not calibrated but the probe function may be checked using calibration solution of known potential. Common solutions contain 3 mmol/L $K_3Fe(CN)_6$ and 3 mmol/L $K_4Fe(CN)_6$ dissolved in 100 mmol/L KCl with

$$E_H(SHE)\,[mV] = 428 + 2.2(25 - t) \tag{4.12}$$

pH buffer solutions saturated with chinhydronium may also be used, with

$$E_H(SHE)\,[mV] = 699 - 59\,pH \qquad t = 25°C \tag{4.13}$$

Common reasons for malfunctions of redox probes are aged electrolyte solutions in the reference electrode, broken cables, surface-contaminated metal electrodes or corroded plugs. In contrast to other potentiometric sensors malfunctions are usually not detected right away since some mV-meters often show stable values which differ well from zero even without a probe connected.

4.4 Materials and Methods

4.4.1 The Artesian Well "Schierensee"

An artesian well in Northern Germany was chosen as a test site for measurements. The well is reported to have discharged oxygen-free water from Tertiary brown coal sands for years. Therefore water samples of homogenous composition could be taken directly without pumping and changes in hydraulics.

Measurements were performed using flow-through cells (KÄSS, 1984) which allow an oxygen-free application of probes. Up to three cells were arranged in serial order. The water flow was adjusted between 0.4 and 0.7 l/min. The flow-through cells were equipped with probes for oxygen, temperature, pH, and with up to six different redox probes.

4.4.2 Electrodes

Four different types of combination electrodes were used (Probes 1 to 4). Additionally, two types of double junction Ag/AgCl reference electrodes were applied in combination with platinum electrodes 5 and 6 (see Table 4.1).

Tab. 4.1: Electrode characteristics.

Probe #	*Type*	*Reference*	*Manufacturer*
1	combination with Pt-tip	$Hg/HgCl_2$	Schott Pt-61
2	combination with Pt-ring	Ag/AgCl	Ingold Pt4805-KN
3	micro-combination with Pt-cone	Ag/AgCl	Ingold Pt4805-M6
4	combination with Pt-ring	Ag/AgCl	Ingold Pt4805-88-NS
5 and 6	massive Pt-tip	(none)	Ingold Pt800-6317
Ref. I	double-junction reference	Ag/AgCl	Orion 90-02-00
Ref. II	double-junction reference	Ag/AgCl	Ingold 373-90-W-NS-S7

4.4.3
Instruments

Oxygen was measured using an *Orbisphere Model 2711* system which is capable of detecting oxygen >50 ppb. pH was measured using a standard *Schott pH 61* probe and a *Knick Portamess 902*. The same instrument was used to measure redox potentials together with a *WTW pH 191*.

4.5
Results

4.5.1
Redox Potential Measurements

Figure 4.3 shows the results of five measurements as difference from the mean measured value. Measurement A was performed using two probes without preparation. Prior to measurement B, all probes were filled with fresh electrolyte solutions and the platinum surface was polished. Measurements C and D were performed without refilling the electrolyte solutions.

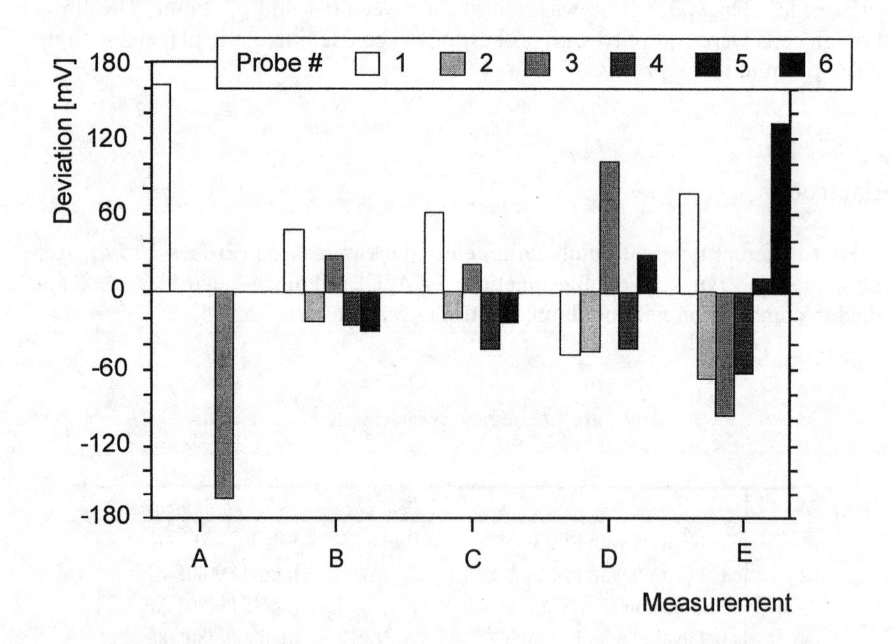

Fig. 4.3: Deviations from mean E_H-values. Results of five measurements at the artesian well "Schierensee" with up to six redox probes (modified from KÖLLING, 1986).

After changing the electrolyte a second time a fifth measurement (E) was performed with all probes. From the results obtained from the measurements B, C and D ageing of the electrolyte solutions was assessed. Yet, differences in measured values became even greater after the probes were refilled prior to measurement E. It should be noted that deviations from the mean are at least ±50 mV even when probes were carefully prepared. Most interesting are the differences observed between probes 5 and 6 which are metal electrodes connected to a separate double-junction reference electrode. We changed the cables the instruments and the reference electrodes and we tried both metal electrodes connected to the same reference. Yet, the difference in potentials between the two identical metal electrodes was still over 100 mV!

Similar deviations from mean values of even more than ±50 mV were observed in sewage water, sea water, tap water and in well waters either saturated or unsaturated with calcite.

4.5.2 Calculation of Redox Potentials

Non-referenced analyses of deep groundwaters were used to compare the calculated redox potentials to the measured values. There was no detailed information on both sampling and measuring conditions, which is normal for many real situations.

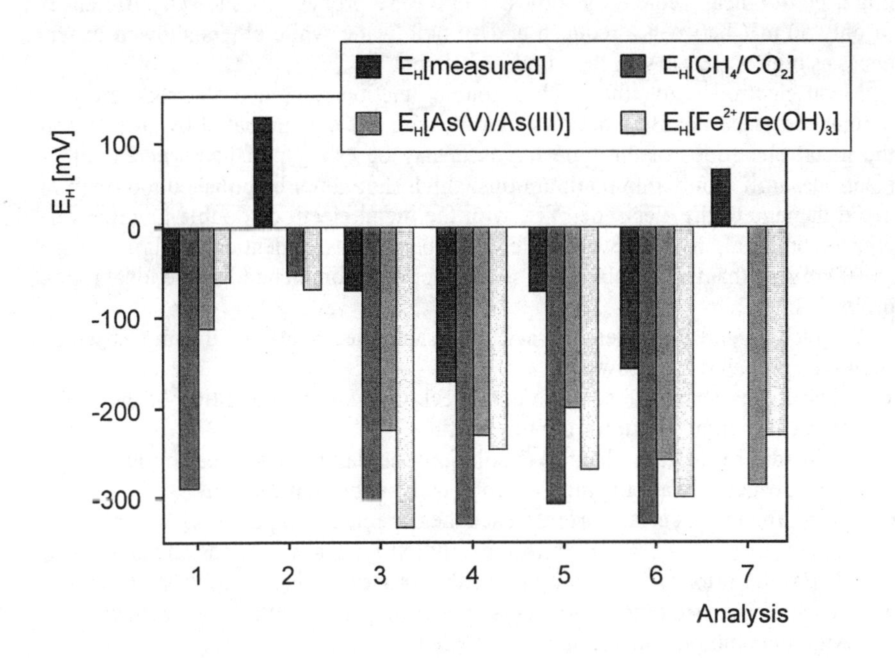

Fig. 4.4: Redox potential calculations for analyses from deep groundwaters (modified from KÖLLING, 1986).

The calculated values are much lower than the measured ones indicating an oxygen contamination of the samples prior to or during measurements. Calculations for the H_2O/O_2 and Fe^{2+}/Fe^{3+}-couples were left out since the results are theoretically much too high as discussed above.

Redox values calculated on the basis of the $Fe^{2+}/Fe(OH)_3$- and As(V)/As(III)-couples differ by less than 100 mV. The absolute values calculated on the basis of the CO_2/CH_4-couple is generally very low due to the low standard potential and the narrow "redox window" (Figure 4.1d). Therefore, agreement with other values is only good when the latter are very low.

4.6 Discussion

4.6.1 Redox Potential Measurements

There are no clear correlations between the probes used and the measurements' accuracy. Great differences among the measured values which might be due to aged electrolyte solution and coated surface of the metal electrode have been suggested, but even with freshly refilled probes and carefully polished electrode surfaces these differences remain. One criterion for the quality of the redox probe might be the measurement dynamics. There were probes that showed differences of only 50 mV between a reduced and an oxic water while others showed differences as high as 260 mV in the same experiment.

Metal electrodes, together with a double-junction reference electrode are recommended since double-junction electrodes are not contaminated by sulfide and the metal electrodes of the type we used may be easily polished whereas other electrodes utilise only thin platinum tips which should not be polished too often to avoid damage to the electrode. Yet, with the metal electrode/double-junction reference combination, there were differences between two identical set-ups as high as 100 mV so that this combination should not be recommended as the ideal redox probe.

We recommend measurements with the same electrode for all samples which should be prepared as follows:

- refill fresh electrolyte solution and check the working condition of the probe with calibration solution from time to time,
- clean and refill fresh electrolyte solution one day prior to measurement campaign, especially after using the probe with calibration solution,
- polish the metal surface prior to each measurement campaign,
- immerse probe in 2% ascorbic acid solution for a few seconds and rinse with pure water prior to every measurement in order to reduce equilibration time.
- Whenever possible, measurements should be performed in flow-through cells without contamination of ambient oxygen.

4.6.2
Redox Potential Calculations

By using the H_2O_2/O_2-couple only a minimum redox potential of a solution may be calculated regardless of the oxygen concentration. A calculation "according to SATO (1960)" is not possible as long as the concentration of the metastable H_2O_2 is not known. Redox potential calculations from the SO_4^{2-}/H_2S-couple are usually faulty due to the slow equilibration of this system and the narrow "redox window", thus showing that the calculations are quite insensitive to measured concentrations. Redox potentials can be determined in natural water with an arsenic content above 10 ppb if species-specific analyses of arsenic are performed. The redox values calculated from the $Fe^{2+}/Fe(OH)_3$-couple usually agree well with the values calculated from As(V)/As(III) and with the measured values as long as the measurements have been performed without oxygen contamination.

4.7
Conclusions

It has to be stated that the redox potential of most natural waters cannot be detected more accurately than ±50 mV. Similar deviations have been determined with two identical metal electrodes connected to one and the same reference electrode. Redox probes have to be carefully prepared to reach this "accuracy". From the As(V)/As(III)-couple and from the Fe^{2+}-concentration of the solution, redox potentials may be calculated at the same "accuracy"-level. This uncertainty of measured redox potentials needs to be carefully considered in thermodynamic calculations.

The determination of redox potentials is usually not interesting in terms of the value itself, but for the geochemical predictions that may be made for the behaviour of a system. It is therefore suitable to measure analytically important redox species as indicators of the redox status of the waters (STUMM, 1984).

4.8
References

CHERRY, J.A.; SHAIK, A.U.; TALLMANN, D.E. & NICHOLSON, R.V. (1979): Arsenic species as an indicator of redox conditions in groundwater. J. Hydrology 43: 373-392.

FREVERT, T. (1984): Can the redox conditions in natural waters be predicted by a single parameter. Schweiz.Z.Hydrol. 46/2: 269-290.

HOSTETTLER, J.D. (1984): Electrode electrons, aqueous electrons and redox potentials in natural waters. Am. J. Science 284: 734-759.

KÄSS, W. (1984): Redoxmessungen im Grundwasser(II). Dt Gewässerkdl. Mitt. 28: 25-27.

KÖLLING, M. (1986): Vergleich verschiedener Methoden zur Bestimmung des Redoxpotentioals natürlicher Wässer. Meyniana 38: 1-19.

PARKHURST, D.L.; THORSTENSON, D.C. & PLUMMER, L.N. (1990): PHREEQE - A Computer Program for Geochemical Calculations. (Conversion and Upgrade of the Prime Version of

PHREEQE to IBM PC-Compatible Systems by J.V. Tirisanni & P.D. Glynn). U.S. Geol. Survey Water Resources Investigations Reports 80-96, Washington D.C.: 195 p.

PARKHURST, D.L. (1995): Users guide to PHREEQC: A computer model for speciation, reaction-path, advective-transport, and inverse geochemical calculations. U.S.Geol.Survey Water Resources Investigations Reports 95-4227, 143 p.

SATO, M. (1960): Oxidation of sulfide ore bodies -1. Geochemical environments in terms of Eh and pH. Econ. Geol. 55: 928-961.

SHAIK, A.U. & TALLMAN, D.E. (1978): Species-specific analyses for nanogram quantities of arsenic in natural waters by arsine generation followed by graphite furnance atomic absorption spectrometry. Anal. Chim. Acta 98: 251-259.

STUMM, W. (1984): Interpretation and measurement of redox intensity in natural waters. Schweiz. Z. Hydrol. 46/2: 291-296.

STUMM, W. & MORGAN, J.J. (1996): Aquatic Chemistry: Chemical equilibria and rates in natural waters. Wiley & Sons, New York, 1022 p.

WHITFIELD, M. (1974): Thermodynamic limitations on the use of the platinum electrode in Eh-measurements. Limnol. Oceanogr. 19: 857-865.

Chapter 5

A Novel Approach to the Presentation of pε/pH-Diagrams

M. Kölling, M. Ebert & H.D. Schulz

5.1 Introduction

Classical pε/pH-diagrams have been constructed assuming an equilibrium between pairs of coexisting species. These conditions yield equations for straight lines where both species show equal activity, separating the pε/pH-diagram into fields in which one or the other species predominates. These diagrams are usually constructed for systems of low complexity, e.g. an aqueous system with fixed element compositions, at fixed pressure and temperature. Usually, both solid phases and dissolved species are presented in one diagram, in which the stability fields for one or more solid phases are shown.

This simplified diagram conceals some important information which is contained in the underlying calculations for its construction. Although we might be interested to learn which aquatic species are predominant in a solution under conditions at which thermodynamics favours the formation of certain minerals, we do not perceive the distribution of dissolved species in this region. The diagram in Figure 5.1 predicts that magnetite is not stable at $pH = 7$ and $p\varepsilon = 10$. Yet, in a superficial neutral oxic environment magnetite will usually not transform to hematite as it possesses long-term stability.

5.2 Construction of pε/pH-Diagrams

We are able to construct diagrams as shown in Figure 5.1 using the appropriate NERNST equations of redox reactions related to the mineral of interest. The diagram in Figure 5.1 actually is composed of two diagrams showing the stability of magnetite and hematite in water, with a total iron activity of 10^{-6} M.

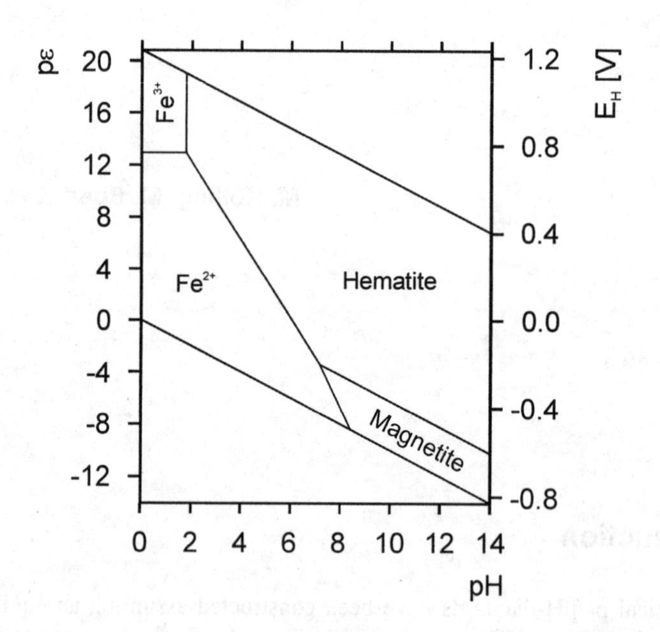

Fig. 5.1: Classical pε/pH composite diagram after GARRELS & CHRIST (1965) showing the stability fields of hematite and magnetite in water. In areas where none of the considered minerals are stable, the fields of dominant dissolved species are shown. The total activity of dissolved iron is 10^{-6} M.

One species distribution visible in this diagram is described by the reaction between Fe^{2+} and Fe^{3+}:

$$Fe^{2+} \leftrightarrow Fe^{3+} + e^-$$

$$p\varepsilon = 13 + \log\left(\frac{[Fe^{3+}]}{[Fe^{2+}]}\right) \tag{5.1}$$

$$p\varepsilon = 13 \quad \text{for } [Fe^{2+}] = [Fe^{3+}]$$

The stability field of hematite is delimited by the reaction with Fe^{3+}:

$$Fe_2O_3 + 6H^+ \leftrightarrow 2Fe^{3+} + 3H_2O$$

$$\log[Fe^{3+}] = -0.72 - 3pH \tag{5.2}$$

$$pH = 1.76 \quad \text{for } \log[Fe^{3+}] = -6$$

and by the reaction with Fe^{2+}:

$$2Fe^{2+} + 3H_2O \leftrightarrow Fe_2O_3 + 6H^+ + 2e^-$$

$$p\varepsilon = 12.34 - \log[Fe^{2+}] - 3pH \tag{5.3}$$

$$p\varepsilon = 18.34 - 3pH \quad \text{for } \log[Fe^{2+}] = -6$$

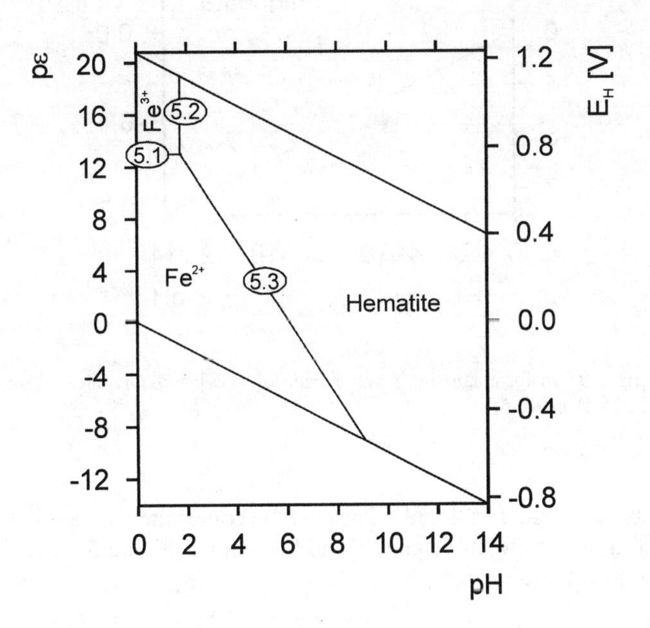

Fig. 5.2: pε/pH-diagram for hematite in water with log $[Fe_T] = -6$. Graphical representation of reactions (5.1), (5.2), and (5.3).

Performing the same calculations for magnetite in water we get:

$$Fe_3O_4 + 8H^+ \leftrightarrow 3Fe^{3+} + 4H_2O + e^-$$

$$p\varepsilon = 5.71 + 3\log[Fe^{3+}] + 8pH \qquad (5.4)$$

$$p\varepsilon = -12.29 + 8pH \quad \text{for } \log[Fe^{3+}] = -6$$

$$2Fe^{2+} + 4H_2O \leftrightarrow Fe_3O_4 + 8H^+ + 2e^-$$

$$p\varepsilon = 16.61 - 1.5\log[Fe^{2+}] - 4pH \qquad (5.5)$$

$$p\varepsilon = 25.61 - 4pH \quad \text{for } \log[Fe^{2+}] = -6$$

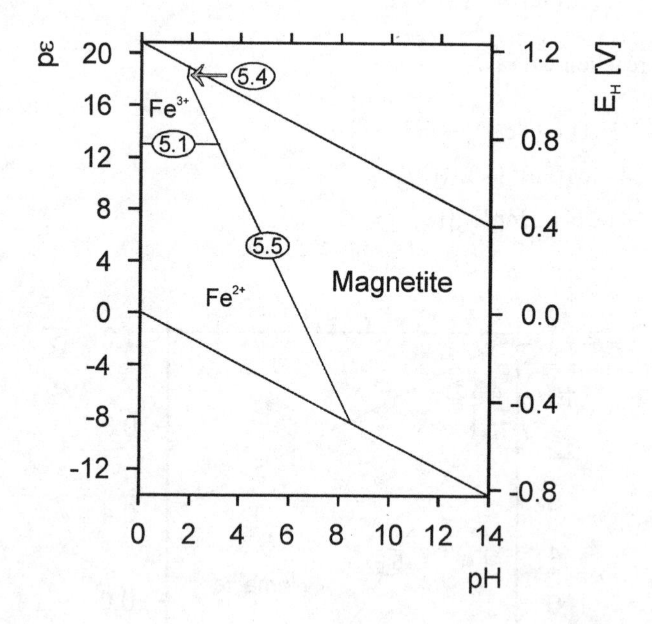

Fig. 5.3: pε/pH-diagram for magnetite in water with log $[Fe_T] = -6$. Graphical representation of reactions (5.1), (5.4) and (5.5).

Superimposing Figures 5.2 and 5.3 we get the composite diagram in Figure 5.4. The delimiting line between magnetite and hematite in Figure 5.1 is equal to line (5.6) in this diagram following

$$p\varepsilon = 3.8 - pH \qquad (5.6)$$

It should be noted that opposed to the classical diagram from GARRELS & CHRIST (1965) there are four fields of mineral stability. There are two fields in which both hematite and magnetite are stable, i.e. both minerals will not be dissolved in water at the given iron activity.

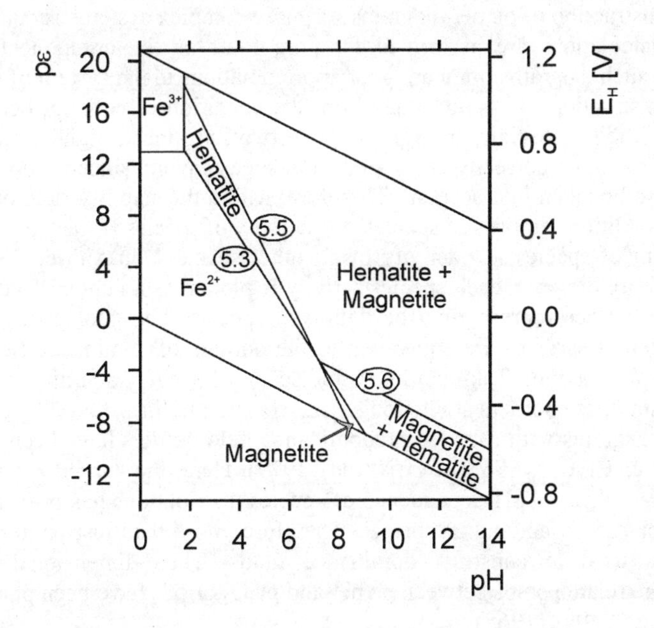

Fig. 5.4: Composite diagram of hematite and magnetite stability in water at log $[Fe_T] = -6$. Graphical representation of reactions (5.3), (5.5) and (5.6).

In the upper field the water is more supersaturated with respect to hematite and therefore magnetite should thermodynamically be transformed into hematite while in the lower field the magnetite saturation is dominant. There are only small triangular fields in the pε/pH-diagram in which only hematite or magnetite are stable in water. Therefore, either one of the minerals will be dissolved.

The difference between the diagrams shown in Figures 5.1 and 5.4 is that mineral stability fields of classical diagrams actually demonstrate mineral saturation predominance fields. Yet, in the major parts of the hematite and magnetite fields shown in Figures 5.1, the concurring mineral is stable as well. If we are interested in interactions between water and minerals rather than in re-crystallisation within mineral parageneses, a representation as in Figure 5.4 is a better tool to assess expected reactions in water.

5.3 Disadvantages of Classical pε/pH-Diagrams

The construction of pε/pH-diagrams for more complex systems is cumbersome. Usually calculations are performed for pure four- or five-component systems. Therefore such diagrams often are poor approximations to the system of interest.

The construction of straight lines from the condition of equality between two species is illicit. Especially in regions where two lines meet, highest activity of a species may not be correctly calculated from species pairs, since all possible species have to be taken into account. Therefore, within the stability field of one species, the condition "activity of species A > activity of species B" has to be replaced by "activity of species A > activity of all other species". In a three-dimensional pε/pH-activity-diagram, each species activity is plotted as a bent surface, with the uppermost surface representing the dominant species. The intersecting lines between different surfaces are equivalent to the borders of dominance fields in the classic pε/pH-diagram. Taking all possible species for every stability field border into account, these borders are no longer strictly straight lines.

pε/pH-diagrams with rounded predominance field borders have been published (GARRELS & CHRIST, 1965; KRAUSKOPF, 1979). Here, the curved borders result from the fact that the pH dependence of species distributions has been taken into account for carbon and sulfur species. Therefore, more than just pairs of species have been used to construct dominance fields. Three-dimensional diagrams showing the relationships between pε/pH-and pCO_2 or pS_2 have been published by GARRELS & CHRIST (1965).

Assuming a constant composition in a solution, constant p-T conditions and a selected set of solid phases, part of the pε/pH-diagram usually shows a stability field for solid phases such that the underlying dominance fields for dissolved species are concealed. In addition, stability fields of solid phases are usually presented as dominance fields where the solid phase with the highest saturation index conceals the underlying stability fields of other solid phases. At the border between two stability fields, the saturation indices of the solid phases are identical.

However, in most natural systems, the phase with the highest saturation index may not be the phase which is actually formed. Other oversaturated phases having a lower saturation index, but a faster rate of precipitation, may form first and control the composition of the solution.

5.4 New Presentation of pε/pH-Diagrams

To circumvent the disadvantages of classical pε/pH-diagrams, we would like to suggest a different method of presentation in which a pair of pε/pH-diagrams is shown, one for the dissolved species and one for the solid phases (Figure 5.5). In one diagram, the dominance fields of dissolved species even is shown in regions where the solid phases are oversaturated.

In the second diagram, overlapping fields are plotted of saturation index SI > 0 for the different minerals. As a result, the stability fields of all solid phases are shown rather than just the field belonging to the phase with the highest saturation index.

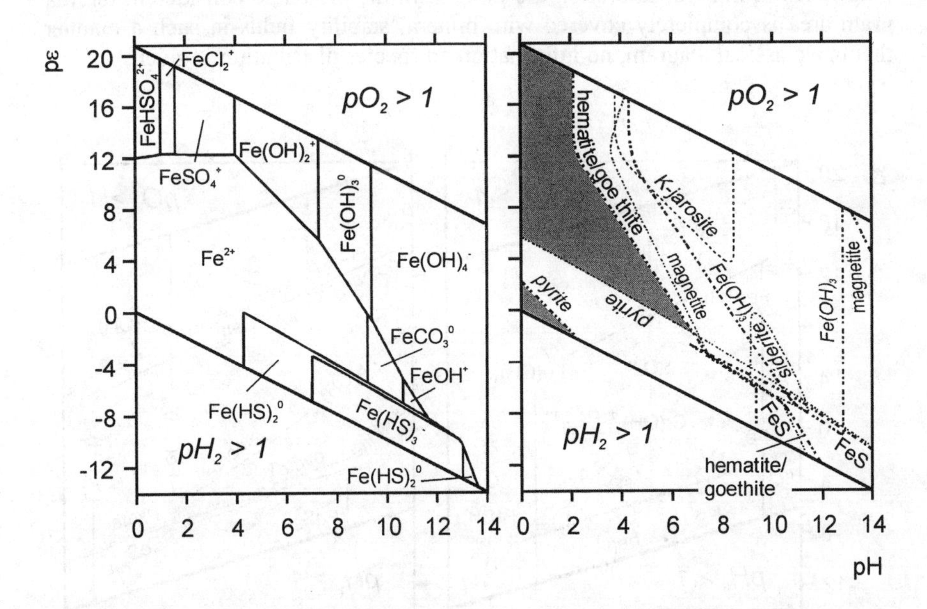

Fig. 5.5: Example for a pair of pε/pH-diagrams for 3 μmol/L Fe in a standard seawater background. *Left:* dominance fields of dissolved iron species. *Right:* overlapping stability fields of different iron minerals. In the shaded area none of the minerals considered are stable. In the classical way of presentation, the dominance fields of the dissolved species are shown only in this area. Note the smooth borders of the stability fields which result from exact calculation of the stability fields accounting for the complete water composition.

To include a greater amount of dissolved species and solid phases, we have performed a series of calculations across the pε/pH-field using thermodynamic model programs such as PHREEQE (PARKHURST et al. 1980, 1990) and PHREEQC (PARKHURST, 1995). Using a pre-processor for the generation of input files and a post-processor for the extraction of activities and saturation indices under different pε/pH-conditions, we were able to construct very precise pε/pH-diagram doublets for both the solid phase stability fields and the dissolved species dominance fields for every system that might be defined in PHREEQE.

Using automated calculation series at a sufficient pε/pH-resolution some interesting curvatures in the stability fields were revealed. Using this method, pε/pH-diagrams may be constructed for real waters rather than for four- or five-component-systems.

EBERT et al. (1997) used this mode of presentation to assess chromate transport in an anoxic FeS-quartz-sand system. In Figure 5.6 a pε/pH-diagram doublet for this system is shown. In the species diagram, iron, sulfur and chromium species distributions are superimposed. It should be noted that the diagram includes the pH-range between pH = 6 and pH = 13. In the mineral stability diagram, overlapping fields of mineral saturation are plotted. In the pH-range considered, the diagram area is completely covered with mineral stability fields in such a manner, that in a classical diagram, no information on species distribution is given.

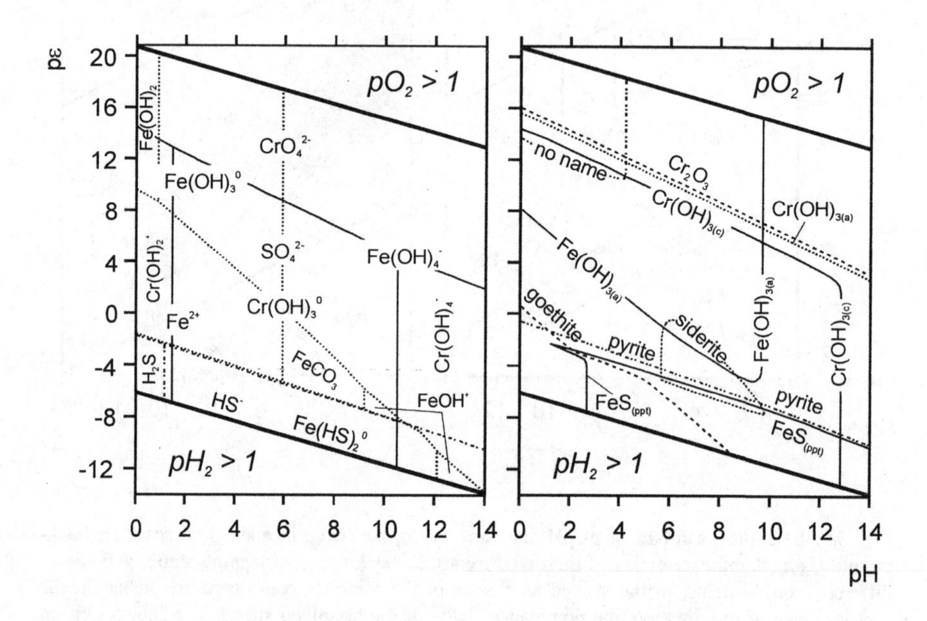

Fig. 5.6: pε/pH-diagram doublet for the assessment of chromate reduction and retention in a slightly alkaline anoxic groundwater system containing FeS (EBERT et al., 1997). The calculations were performed using PHREEQE for Cr (0.5 mM), Fe (1 μM) and S (0.5 mM) in a real water with Ca (0.26 mM), Mg (0 15 mM), Na (1.9 mM), K (0.1 mM), Mn (3 μM), Cl (0.63 mM), N(0.4 mM), P (3.1 μM), and IC (0.92 mM). The data scheme is 4 points per pε and 4 points per pH i.e. this diagram contains data from approximately 2000 PHREEQE runs. Methane formation was neglected. The mineral „no name" is the jarosite-type $KFe_3(CrO_4)_2(OH)_6$.

The diagrams show that chromate does not coexist with dissolved sulfide or ferrous iron species, nor with the sulfide minerals or the ferrous iron minerals under study. Although the reactions with sulfides should be thermodynamically favoured due to these diagrams, chromate reduction by sulfide only occurs at high chromate levels that inhibit iron reducing bacteria.

5.5 Conclusions

Upon investigating dissolution/precipitation reactions, the composition of the solution in contact with a mineral which, due to oversaturation, is likely to form is of major interest. Using two separate diagrams, one for dissolved species and one for solid phases, opens the view into the regions that are concealed by mineral stability fields in classical diagrams.

Having to assess, whether or not a natural water is aggressive to the mineral of interest, or whether the formation of a certain mineral is thermodynamically possible, although it might not represent the thermodynamically stable phase in the long term, plotting overlapping fields of mineral stability is much more useful than classical plotting of mineral predominance fields.

By applying several series of PHREEQE calculations to any particular pε/pH-range, very precise diagram doublets can be constructed even applied to complex real systems

5.6 References

EBERT, M.; ISENBECK-SCHRÖTER, M. & KÖLLING, M. (1997): Reduction and retention of chromate in an anoxic quartz-sand-FeS-system – A laboratory study in water-saturated columns.- In: EBERT, M.(1997): Der Einfluß des Redox-Milieus auf die Mobilität von Chrom im durchströmten Aquifer.- Ber. FB Geowiss. Univ. Bremen 101: 70–99.

GARRELS, R.M. & CHRIST, C.L. (1965): Solutions, Minerals, and Equilibria.- Harper & Row, New York, 450p.

KRAUSKOPF, K.B.(1979): Introduction to Geochemistry.- 2nd ed. McGraw Hill, New York; 617p.

PARKHURST, D.L.; THORSTENSON, D.C. & PLUMMER, L.N. (1980): PHREEQE - A Computer Program for Geochemical Calculations. - U.S. Geol. Survey Water Resources Investigations Reports 80-96, Washington D.C.: 210p.

PARKHURST, D.L.; THORSTENSON, D.C. & PLUMMER, L.N. (1990): PHREEQE - A Computer Program for Geochemical Calculations. (Conversion and Upgrade of the Prime Version of PHREEQE to IBM PC-Compatible Systems by J.V. Tirisanni & P.D. Glynn).- U.S. Geol. Survey Water Resources Investigations Reports 80-96, Washington D.C.: 195p.

PARKHURST, D.L. (1995): Users guide to PHREEQC: A computer model for speciation, reaction-path, advective-transport, and inverse geochemical calculations.- U.S.Geol.Survey Water Resources Investigations Reports 95-4227, 143p.

Chapter 6

The Couple As(V) - As(III) as a Redox Indicator

T. R. Rüde & S. Wohnlich

6.1 Introduction

The main focus of this textbook is on redox measurements using Pt-electrodes. The following chapter is about the calculation of E_H or pε using analytical data of redox couples instead of direct measurements with electrodes. This approach is immediately derived from the NERNST equation as discussed in Chapter 1 which relates the redox potential to the ratio of an oxidised to a reduced species, i.e.

$$E_H = E_H^0 + \frac{RT}{nF} \ln \frac{\{Ox\}^o}{\{Red\}^r} \qquad (6.1)$$

where:

E_H^0 = standard potential [mV]
T = absolute temperature [K]
R = gas constant ($8.314 \cdot 10^{-3}$ kJ·K^{-1}mol^{-1})
n = number of electrons transferred in the reaction
F = Faraday´s constant (96484.56 C/mol),

{Ox}, {Red} = activities of the oxidised and reduced species with their stoichiometric coefficients o and r.

Instead of calculating the E_H by the NERNST equation it is also very popular to use the concept of the hypothetical electron activity, i.e. the pε-concept. The equation of a half cell in logarithmic form is

$$p\varepsilon \quad p\{Re\,d\} - p\{Ox\} + pK \tag{6.2}$$

where pK denotes the negative logarithm to base 10 of the equilibrium constant of any redox reaction. This concept is straightforward to calculations of dissociation reactions, complex formation, dissolution-precipitation or sorption equilibria. Both parameters, i.e. E_H and pε, are related to each other

$$E_H = \frac{2.303\,RT}{F}\,p\varepsilon \tag{6.3}$$

This approach can be considered as an obvious method that can compete with electrode measurements especially as the latter method contains problems which are discussed in other chapters of this book besides the advantage of in-situ measurements.

Any discussion of the approach using the NERNST equation should regard that this is not a simple method of redox estimation and it is the aim of the following pages to show every single step between the first sampling and the final value of the redox potential. Theoretically, there are several species couples which would allow an estimation of the redox status of a natural water. Couples like Fe(III)/(II), Mn(IV)/(II), SO_4^{2-}/HS^-, NO_3^-/NH_4^+ or CH_2O/CH_4 are typically shown on diagrams like the one given in Figure 1.2 of this book describing the evolution of environmental systems from oxidised to increasingly reduced conditions. But it should be mentioned that the use of such couples has some limitations in practice.

This discussion will use the arsenic redox couple as an example. The use of this couple is dating back to the refreshing work of CHERRY and his co-workers (1979) who first discussed the thermodynamic properties of this redox couple in relation to the use as a redox indicator in groundwaters.

6.2 The Procedure

The calculation of the redox status by an ion couple consists of several steps which are the theme of the discussion in this chapter:

- sampling
- analysing As(III) and As(V)
- analysing the main ions
- calculating the ionic strength
- calculating activity coefficients

- calculating the activities of the As species
- drawing up a system of thermodynamic redox equations
- solving the thermodynamic equations to get pε-values.

The last two points give an indication to a certain notion of how to approach redox couples and should be kept in mind. The procedure relies on a thermodynamic equilibrium and can only result in meaningful results if the conditions of the natural environment are not too far from a steady state.

6.2.1 Sampling

The oxidation of arsenic(III) by oxygen is a relatively slow process (BÖKELEN & NIESSNER, 1992; CHERRY et al., 1979; TALLMAN & SHAIKH, 1980) in comparison to the oxidation of e.g. reduced sulfur species or nitrite. This allows to handle samples for arsenic(III) determinations outside glove boxes or similar protection systems. However, there are some matrix constituents which can cause serious problems leading to lower or even higher (RÜDE, 1996; STUMMEYER et al., 1995) As(III) concentrations in samples. Especially the last case should not be underestimated, as arsenate(V) can be totally reduced to As(III) in the microenvironment of sample containers (e. g. RÜDE, 1996). Measurements of As(III) are in no instance measurements of least concentrations.

Most serious matrix constituents in natural samples are sulfur hydrogen enhancing As(III) concentrations and Fe(III) reducing As(III) concentrations (CHERRY et al., 1979, TALLMAN & SHAIKH, 1980). Due to this, samples for arsenic species analysis should be measured as fast as possible, but it is possible to increase the critical time span to some days or even weeks by employing some measures of preservation. CHEAM & AGEMIAN (1980) found that an acidification of samples with 0.2 % (v/v) of sulfuric acid allows natural surface waters to be stored at room temperature for 125 days. AGGETT & KRIEGMAN (1987) stored samples of sedimentary interstitial waters after acidification to pH 2 with HCl and cooling to 2 °C for over 6 weeks with no change in arsenic speciation. RÜDE (1996) found that storage of thermal waters at a temperature of 2 °C is sufficient to stabilise the arsenic species for ten days. In summary, merely cooling seems to be sufficient when the samples are low in iron (< 100 μg/L), but for samples with high iron concentrations acidification will be required in addition to cooling.

Cooling of samples will also reduce the activity of micro-organisms. A variety of micro-organisms are capable of altering the concentrations of inorganic arsenic species by bio-methylation and de-methylation (CUTTER et al., 1991; HASWELL et al., 1985; LÉONARD, 1991). Furthermore, the influence of microbial processes on other ions, e.g. microbial sulfate reduction to hydrogen sulfide, can seriously effect the speciation of arsenic. In the very case, the influence of micro-organisms has to be investigated.

Regarding the material of sample containers, most of the commonly used materials are suitable for arsenic samples. CHEAM & AGEMIAN (1980) found no difference between glass and polyethylene bottles. In case samples need deoxygenation, AGGETT & KRIEGMAN (1987) recommended glass bottles instead of polyethylene containers. RÜDE (1996) found no statistically significant difference between containers out of glass (two qualities), polyethylene, polypropylene or fluorinated ethylene propylene storing thermal waters at 2° C for 10 days. Although, it seems that the storage in glass or denser plastics is preferable to polyethylene.

6.2.2 Analysis

It is out of the scope of this chapter to discuss the various procedures to determine the arsenic speciation in detail. In the past two decades applications have been developed for most analytical techniques. To give an idea of the huge literature on this theme just a view publications should be mentioned, e.g. methods using ion chromatography (AGGETT & KADWANI, 1983; GRABINSKI, 1981; IVERSON et al., 1979), high performance liquid chromatography (EBDON et al., 1988; HANSEN et al., 1992), gas chromatography (CUTTER et al., 1991; ODANAKA et al., 1985), liquid-liquid-extraction (HUANG & WAI, 1991), polarography (CHAKRABORTI et al., 1984), cold traps (ANDREAE, 1977; BRAMAN et al., 1977; MASSCHELEYEN et al., 1991) and species sensitive hydride generation techniques (AGGETT & ASPELL, 1976; ANDERSON et al., 1986; GLAUBIG & GOLDBERG, 1988; WEIGERT & SAPPL 1983). These citations are not a representative list of all works done on techniques to measure arsenic species.

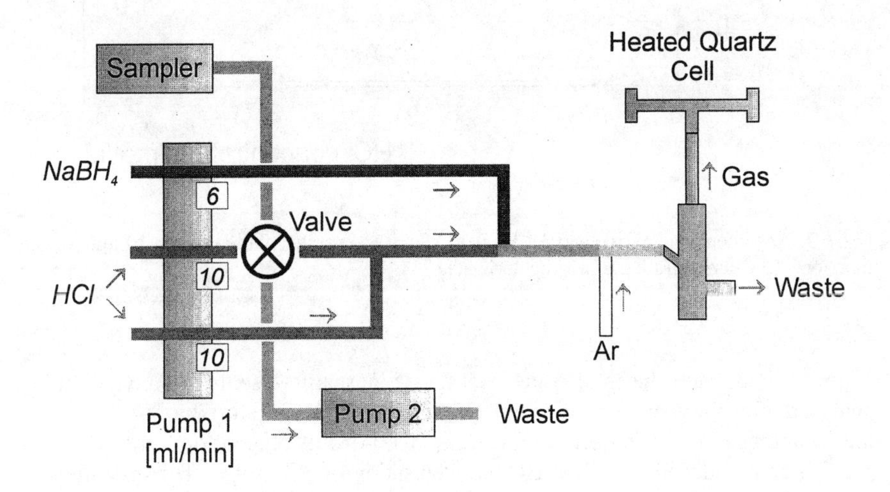

Fig. 6.1: Schematic drawing of the apparatus used for As(III) sensitive hydride generation technique. The sample is injected with a 4-way valve and thoroughly mixed with hydrochloric acid before reduction by $NaBH_4$ takes place.

The data given here were derived by species sensitive hydride generation (RÜDE, 1996). The basics of this method will be sketched just to give the complete background how the redox data at the end of the chapter were produced. It should not be understood that this method is preferable to other ones.

Using a commercially available flow injection system (Perkin-Elmer FIAS 400), the sample is injected with a valve into an acid stream (HCl) and after thoroughly mixing reduced by a stream of sodium tetrahydridoborate ($NaBH_4$). Arsenic is then quantified by atomic absorption in a heated quartz cell after the separation of the gaseous arsenic hydride from the liquid sample matrix (Figure 6.1). As is shown on Figure 6.2, the choice of the molarity of the acid allows to suppress the reduction of arsenic(V) in favour of arsenic(III). One should work at an acid concentration above 3 mol/L to receive also a good sensitivity for arsenic(III). A concentration of 4.0 mol/L of HCl was used for the analyses discussed in this chapter.

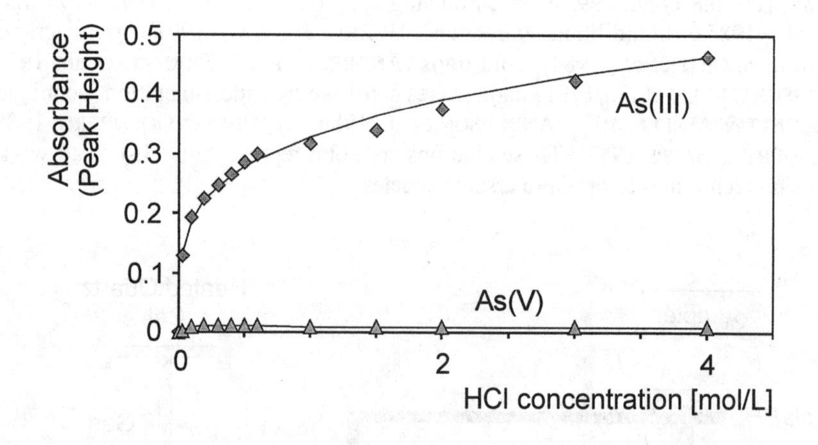

Fig. 6.2: Absorbency of As(III) and As(V) at 193.7 nm – separate solutions, each 10 µg/L – as a function of the concentration of hydrochloric acid.

In a second step the total concentration of inorganic arsenic is conventionally analysed after the reduction of arsenic(V) to arsenic(III) by a mixture of potassium iodide and ascorbic acid each 10 g/L using 1 ml of this reagent on 9 ml of a sample solution acidified to a final HCl concentration of 2.4 mol/L. This treatment is done in the sample cups and 45 minutes should be allowed for a complete reduction of the arsenic(V). The concentration of As(V) can then be calculated as the difference between the total concentration of inorganic As and As(III).

It should be mentioned that the determination of arsenic in 4.0 mol/L HCl only suppresses signals of As(V) but does not totally forbid for arsane generation from As(V). In the case of very high As(V) to As(III) ratios, the outcome of the 4.0 mol/L HCl determination is a composite value which has to be corrected with regard to As(V) contributions. This can be done by a simple iteration, but nonetheless, this is the first calculation procedure contributing to the final figure of pε.

6.2.3
Analysing the Major Ions

It is indispensable for a redox calculation to analyse also the major ions of a water sample. Regarding this prerequisite, it is obvious that the approach using redox sensitive species couples is not a fast screening method. The necessary data base is only available after detailed investigations, during which the necessary analyses will be carried out. The standard analytical methods will not be discussed here.

The following part of this chapter uses samples from research work in the former open pit Cospuden in the Eastern German lignite mining district south of the city of Leipzig for the process of redox calculation with the arsenic species couple.

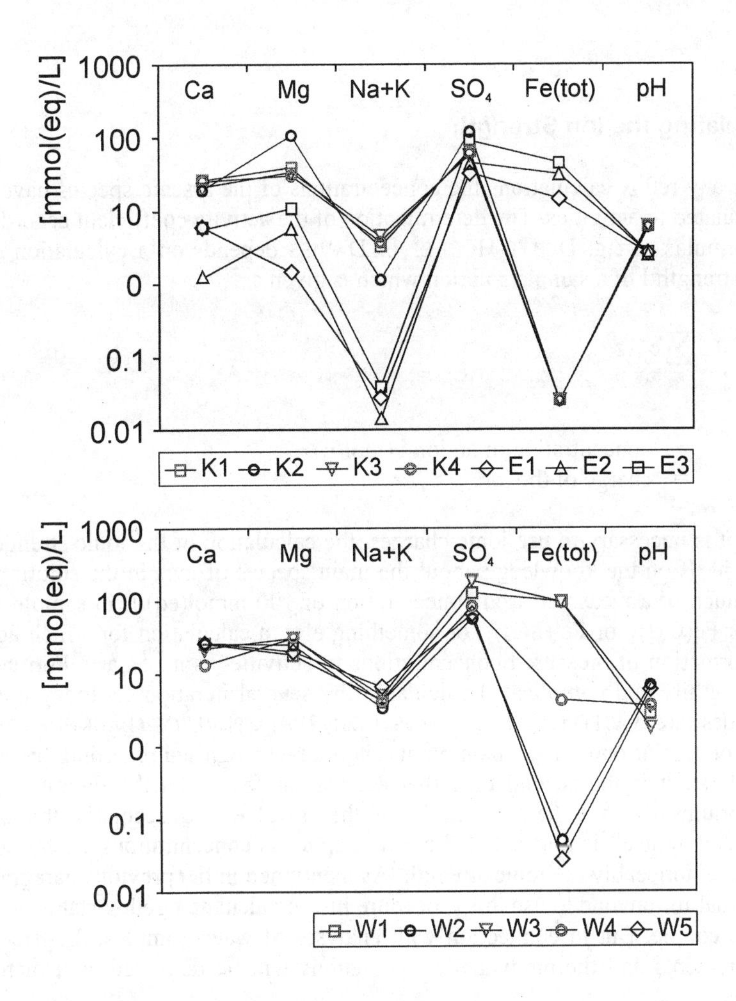

Fig. 6.3: Main composition of samples of the spoil dump of the former open pit Cospuden. *K* denotes samples collected with suction cups, *E* water extracts of spoil material and *W* seepage water sampled at the top of the spoil dump.

There are three sample types shown on the semilogarithmic diagrams of Figure 6.3, which are based on the diagram developed by SCHOELLER (1962). Samples marked by *K* are seepage collected with suction cups, group *E* are extracts from batch experiments using water to extract samples of the research borehole Cospuden (for a complete description see WOHNLICH et al., 1997) drilled into the spoil dump, and the third group marked by *W* refers to seepage water emerging at the surface of the dump. Main features of these samples are their acidic pH-values, the dominance of Mg^{2+} over Ca^{2+}, the relatively low concentrations of alkaline elements, the high sulfate concentrations and the bias in iron concentrations between samples with high values between 10 and 100 mmol(eq)/L and samples with low values around 0.05 mmol(eq)/L. All these samples are characterised by products of sulfide oxidation and set up a collection of samples which are so high in total dissolved solutes (TDS) that it is unreasonable to set concentrations equal to activities.

6.2.4
Calculating the Ion Strength

For any redox calculations the concentrations of the arsenic species have to be recalculated to activities. The determination of the activity coefficient according to the formulas of e.g. DEBYE-HÜCKEL or DAVIES depends on a calculation of the ionic strength I of a sample solution which is given as

$$I = 0.5 \sum c_i \cdot z_i^2 \tag{6.4}$$

where:

c = concentration of an ion i (mol/L)
z = charge of the ion.

As it is necessary to use ionic charges, the calculation of the ionic strength depends itself on the knowledge about the main species of ions in the solution, e.g. how much of a measured iron concentration of 100 mmol(eq)/L in sample *W1* is Fe^{3+} or $Fe(OH)_2^-$ or $Fe_2(SO_4)_3^0$ or something else, a calculation for which again a transformation of measured concentrations to activities is necessary. Fortunately, this lengthily and complicated calculation by several iterations is today done by PC-codes like WATEQ (BALL & NORDSTROM, 1991), PHREEQE (PARKHURST et al., 1980) or similar ones. Two main points should be recognised regarding the procedure. First, it is the second time that calculations based on the thermodynamic equilibrium influence the final result, i.e. the pε-value, and, secondly, the arsenic redox couple itself is not included in this step as its concentrations are too low to influence noticeably the ionic strength. As mentioned in the previous paragraph, it is unusual for anyone to use this procedure just to calculate a redox status with the arsenic couple. But in connection with analyses of water samples, the necessary measurements and thermodynamic calculations will be delivered as a matter of routine.

6.2.5 Calculating the Activity Coefficients

The remaining steps in the whole procedure delivering redox values will focus on the arsenic species themselves. Using the calculated ionic strengths it is possible to set up the activity coefficients of the arsenic species either by the extended DEBYE-HÜCKEL equation up to ionic strengths of 0.1:

$$\lg \gamma_i = -Az_i^2 \frac{\sqrt{I}}{1 + Ba\sqrt{I}} \tag{6.5}$$

or the DAVIES equation for ionic strengths < 0.5:

$$\lg \gamma_i = -Az_i^2 \left(\frac{\sqrt{I}}{1 + \sqrt{I}} - 0.3\,I \right) \tag{6.6}$$

where:

A and B = temperature depending coefficients

a = a parameter corresponding to the size of the ion *i*.

Tab. 6.1: Calculated activity coefficients (DAVIES equation) of the samples from Cospuden.

Sample	ionic strength *DAVIES*	*Activity Coefficient* Charge 1	Charge 2	Charge 3
K1	0.09	0.94	0.77	0.56
K2	0.15	0.91	0.67	0.41
K3	0.08	0.94	0.79	0.59
K4	0.08	0.94	0.79	0.59
W1	0.14	0.91	0.69	0.43
W2	0.08	0.94	0.80	0.60
W3	0.19	0.89	0.63	0.35
W4	0.09	0.94	0.77	0.56
W5	0.08	0.94	0.79	0.59
E1	0.03	0.97	0.90	0.79
E2	0.04	0.97	0.88	0.75
E3	0.08	0.94	0.78	0.58

Table 6.1 shows the calculated activity coefficients γ for the three negatively charged species of the arsenic acid. The undissociated acid itself and the arsenious acid have by definition an activity coefficient of 1. As in most natural waters, the species of As(III) is the undissociated arsenious acid H_3AsO_3. The ionic strengths and the activity coefficients are calculated using the DAVIES equation.

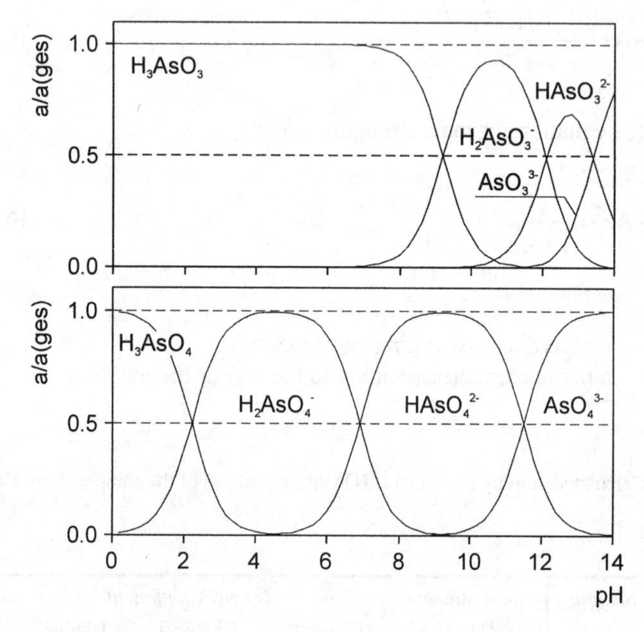

Fig. 6.4: Dissociation diagrams of arsenious acid (top) and arsenic acid (bottom), where a denotes the activities of the arsenic species.

6.2.6 Calculating the Activities of the As-Species

The transformation of the measured As concentrations to activities using the activities coefficients raises a further problem, i.e. the question of the amount of certain species contributing to the measured concentrations. This situation is quite simple for As(III) as H_3AsO_3 is the only species of As(III) in most natural waters. In contrast to that, it is to be expected that two species are mainly contributing to an As(V)-value, e.g. H_3AsO_4 and $H_2AsO_4^-$ or $H_2AsO_4^-$ and $HAsO_4^{2-}$ which require different activity coefficients. To illustrate this, Figure 6.4 depicts dissociation diagrams of arsenious and arsenic acid. The curves are calculated by using the following algorithm (BLIEFERT, 1979):

$$\left\{H_{n-1}A^{i-}\right\} = \frac{\left\{H_3O^+\right\}^{n-i} \cdot \left\{H_nA\right\} \cdot \prod_{j=1}^{i} K_j}{\sum_{r=0}^{n}\left(\left\{H_3O^+\right\} \cdot \prod_{j=1}^{r} K_j\right)} \qquad (i = 1, 2, \ldots, n) \qquad (6.7)$$

where the brackets denote activities of dissociation species i. The dissociation constants are taken from the compendium of NAUMOV et al. (1974).

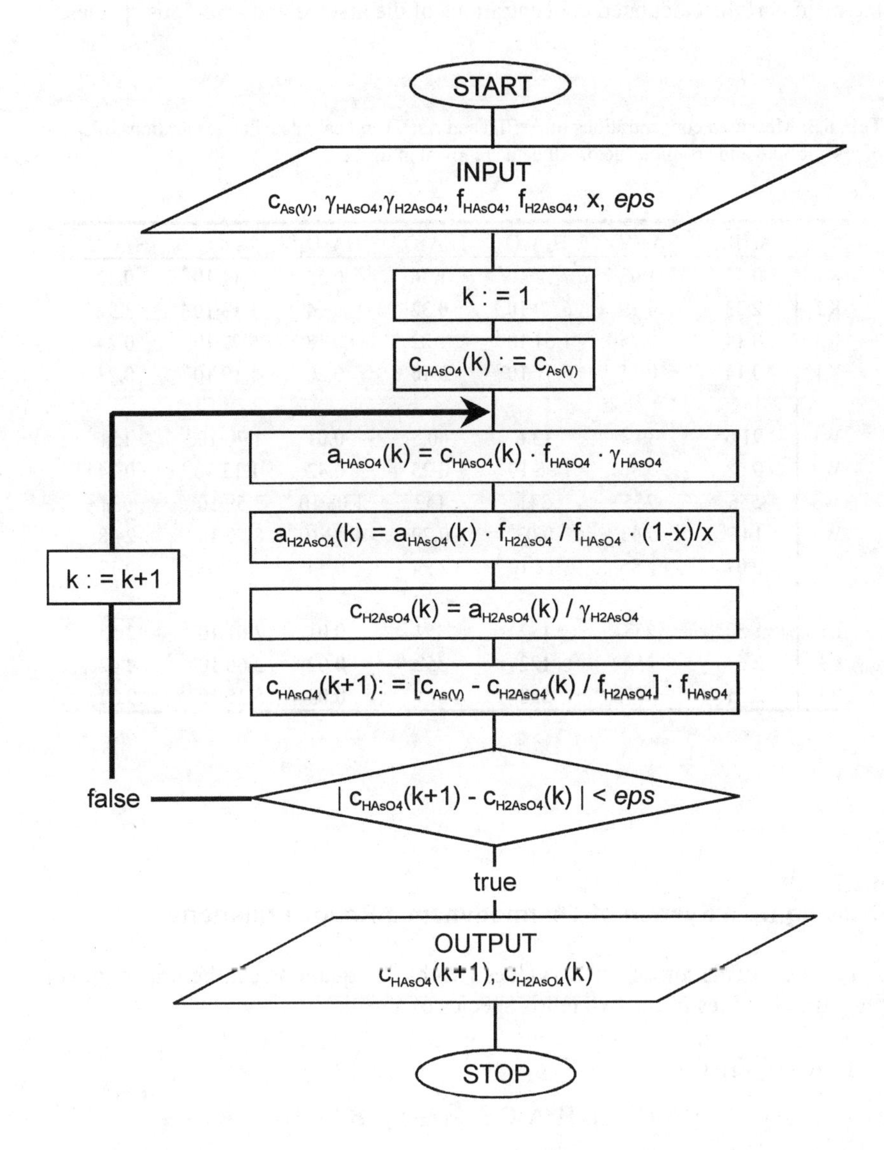

Fig. 6.5: Iteration scheme used to calculate the concentrations of different As(V) species, where c is concentration, a is activity and f are transformation factors for different molecules.

Again - for the third time - iterative calculations are necessary to solve the problem. The code used to calculate the As(V) species of the samples of Cospuden is shown in Figure 6.5 simplified to two species. As the value of x, i.e. the contribution of a certain species to the total As(V) activity, is taken from the dissociation algorithm, the question may arise why it is not directly used to calculate the amount of each As(V) species. Here it should be considered that the ionic strength of the sample solutions forbids to set concentrations and activities equal. Table 6.2 gives the results of these calculations for the chosen examples from Cospuden depicting on the left side the measured concentrations of As(III) and As(V) and in the main part the calculated concentrations of the arsenic and arsenious species.

Tab. 6.2: Measured concentrations of As(III) and As(V) and calculated concentrations of species of arsenic acid and arsenious acid. All data are given in µg/L.

	As(III)	As(V)	H_3AsO_4	$H_2AsO_4^-$	$HAsO_4^{2-}$	AsO_4^{3-}	H_3AsO_3
K1	0.13	1.02	$7.52 \cdot 10^{-6}$	1.36	0.55	$1.11 \cdot 10^{-5}$	0.22
K2	2.33	0.19	$6.50 \cdot 10^{-5}$	0.32	0.04	$1.13 \cdot 10^{-7}$	3.92
K3	0.14	0.75	$1.64 \cdot 10^{-4}$	1.22	0.18	$6.92 \cdot 10^{-7}$	0.24
K4	0.14	0.51	$5.41 \cdot 10^{-5}$	0.76	0.20	$1.19 \cdot 10^{-6}$	0.23
W1	91.6	39.4	13.6	60.5	0.01	$1.70 \cdot 10^{-11}$	154
W2	0.08	0.36	$1.96 \cdot 10^{-6}$	0.25	0.42	$1.13 \cdot 10^{-5}$	0.14
W3	556	255	341	142	$1.08 \cdot 10^{-3}$	$3.59 \cdot 10^{-13}$	934
W4	14.7	0.43	0.03	0.79	$4.79 \cdot 10^{-4}$	$8.70 \cdot 10^{-12}$	24.8
W5	0.07	1.85	$3.12 \cdot 10^{-4}$	2.94	0.54	$2.42 \cdot 10^{-6}$	0.12
E1	1.50	253	132	354	0.01	$2.01 \cdot 10^{-11}$	2.53
E3	2.63	252	122	352	0.02	$2.66 \cdot 10^{-11}$	4.42
E3	5.55	477	238	661	0.03	$5.35 \cdot 10^{-11}$	9.32

6.2.7 Drawing Up a System of Thermodynamic Redox Equations

For the calculation of the pε-values two basic equations can be drawn up coupling the activities of the two redox species of arsenic

$$\begin{aligned} &H_3AsO_3 + H_2O \leftrightarrow \\ &\quad x \cdot H_3AsO_4 + (1-x) \cdot H_2AsO_4^- + (3-x) \cdot H^+ + 2e^- \end{aligned} \tag{6.8}$$

$$
\begin{aligned}
&H_3AsO_3 + H_2O \leftrightarrow \\
&\quad x \cdot H_2AsO_4^{-} + (1-x) \cdot HAsO_4^{2-} + (4-x) \cdot H^{+} + 2e^{-}
\end{aligned}
\tag{6.9}
$$

The laws of mass action of these equations give in logarithmic form and rearranged for pε

$$
\begin{aligned}
p\varepsilon = 0.5(&pK + x\lg\{H_3AsO_4\} + (1-x)\lg\{H_2AsO_4^{-}\} \\
&- \lg\{H_3AsO_3\} - (3-x)pH)
\end{aligned}
\tag{6.10}
$$

$$
\begin{aligned}
p\varepsilon = 0.5(&pK + x\lg\{H_2AsO_4^{-}\} + (1-x)\lg\{HAsO_4^{2-}\} \\
&- \lg\{H_3AsO_3\} - (4-x)pH)
\end{aligned}
\tag{6.11}
$$

The dissociation constants for other than standard conditions, i.e. temperature deviating from 298.16 K, can be calculated by using the VAN'T HOFF equation, experimentally determined algorithms as e.g. published by SPYCHER & REED (1989) or using under special prerequisites standard free (GIBBS) energies

$$
\Delta G_r^0 = -RT\ln K \tag{6.12}
$$

In the special case of the samples discussed as examples herein the latter approach is feasible as the actual temperatures of the samples are only slightly different from 298.16 K - the whole range of the samples is 292.36–295.76 K - and no gas phases are involved in the reactions. In logarithmic form and rearranged for pK the equations are:

$$
\begin{aligned}
pK = \frac{1}{2.303RT}(&x\Delta G_f\{H_3AsO_4\} + (1-x)\Delta G_f\{H_2AsO_4^{-}\} \\
&- \Delta G_f\{H_3AsO_3\} - \Delta G_f\{H_2O\})
\end{aligned}
\tag{6.13}
$$

$$
\begin{aligned}
pK = \frac{1}{2.303RT}(&x\Delta G_f\{H_2AsO_4^{-}\} + (1-x)\Delta G_f\{HAsO_4^{2-}\} \\
&- \Delta G_f\{H_3AsO_3\} - \Delta G_f\{H_2O\})
\end{aligned}
\tag{6.14}
$$

This set of six equations will finally allow to calculate pε-values of the water samples. But a last obstacle should be mentioned, i.e. the choice of accurate thermodynamic data, in this case the standard free energies.

A review of the literature on thermodynamic data of arsenic reveals at a first glance a surprising bias regarding the dissociation constant of the second dissociation step of the arsenious acid. BROOKINS (1988) in his generally quite useful book shows a value around $pK_2 = 11$ which is adopted also by VINK (1996) in his re-evaluation of the Sb- and As-diagrams. In contrast to that, e.g. CHERRY et al. (1979) and RAI et al. (1984) give a value around 12 which is also cited in the database of MINTEQA2 (BROWN & ALLISON, 1987). The reason for this difference

Tab. 6.3: Free (GIBBS) energies for arsenic species (Data taken from NAUMOV et al. (1974), originally given in [kcal/mol] and recalculated to [kJ/mol] by multiplying with 4.184).

As(III)	$\Delta G^0{}_f$ [kJ/mol]	As(V)	$\Delta G^0{}_f$ [kJ/mol]
H_3AsO_3	-646.01	H_3AsO_4	-770.02
$H_2AsO_3^-$	-593.33	$H_2AsO_4^-$	-757.47
$HAsO_3^{2-}$	-524.4	$HAsO_4^{2-}$	-717.68
AsO_3^{3-}	-447.7	AsO_4^{3-}	-651.99

seems to be a division in the data used by BROOKINS (1988). He used for aqueous arsenic species the NBS data base (WAGEMAN et al., 1982) but cited DOVE & RIMSTIDT (1985) for $HAsO_3^{2-}$ and AsO_3^{3-} as WAGEMAN and his co-workers are assuming arsenious acid in its metaform $HAsO_2$. This way of citation is confusing as DOVE & RIMSTIDT (1985) themselves are citing NAUMOV et al. (1974), i.e. the American publication of the 1971 published data of the former *Soviet Academy of Sciences*.

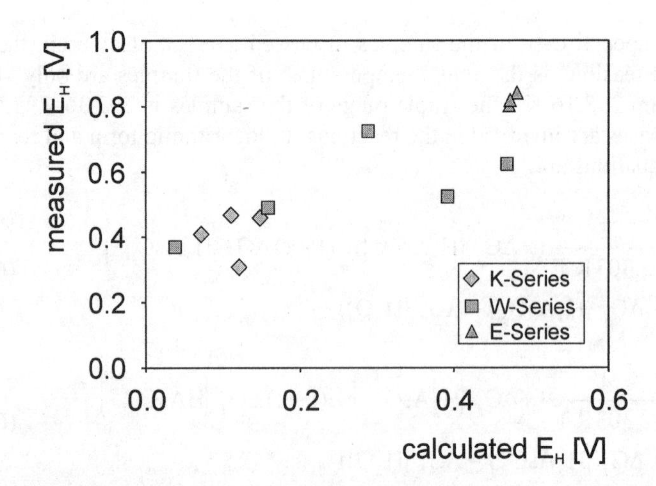

Fig. 6.6: Comparison of E_H-values of the Cospuden samples measured by a platinum electrode and calculated by the As(III) - As(V) redox couple.

It is obvious that thermodynamic data of arsenic are mainly derived from two sources: the former *National Bureau of Standards* (WAGEMAN et al., 1982) and the *Soviet Academy of Sciences* (NAUMOV et al., 1974). The Soviet data are systematically lower than the values of the NBS. Although calculations based on only one data set will give consistent and comparable results, a mixing of both data sets will lead to false results at the interface between both sets, i.e. a false calculation of the second dissociation constant of the arsenious acid. This example demonstrates that thermodynamic data should be chosen carefully and that it is in any case unwise to mix up different data sets.

The calculations done in the context of this chapter are all based on the compilation of NAUMOV et al. (1974) and for convenience the data are summarised in Table 6.3.

6.2.8 Solving the Thermodynamic Equations to get pε

After all these steps of analyses and calculations the final results are pε-values which can easily be transformed into E_H-values. For the example of Cospuden the results of this procedure are compared in Figure 6.6 with data measured by a platinum electrode (Ingold Pt 4805-S7, plugged to a WTW voltmeter). The calculations lead to lower values with some trend between both methods although the number of data is too small to compare both methods statistically. But these data do not scatter as much as the data published by LINDBERG & RUNNELLS (1984), whose diagram is shown in Figure 1.4.

6.3 Conclusions

The use of redox sensitive ion couples as a tool to measure the redox status of a natural system is by theory a straightforward approach which is based on the NERNST equation, one of the most principal equations of electrochemistry. CHERRY et al. (1979) have shown in their basic work that especially the As(III) - As(V) - couple is an useful tool in this approach as it is responding to the natural pε-conditions fast enough to give an indication of them. On the other hand the readjustment to new conditions is slow enough to preserve a certain natural status in a sample long enough to be measured. Furthermore, in the past two decades the analytical techniques to determine the arsenic speciation have made great progress. Nowadays methods to measure the speciation of arsenic are developed for almost all of the major analytical techniques.

The aim of this chapter was to delineate the single steps that are necessary to use this tool for measuring even waters high in TDS, i.e. where a recalculation of concentrations to activities becomes indispensable. Such a discussion reveals that this process relies on e.g. up to three iterative calculations and also a careful selection of thermodynamic data. All the mathematical steps here emphasised can be installed in simple PC programs to be handled in a black box manner, i.e. just to plot the analytical data into the program and to receive a final pε or E_H-value.

Regarding one of the most basic analytical practices, 15 years ago LINDER et al. (1984) mentioned that "a pH measurement is definitely not just a matter of switching on a pH meter, plunging the electrodes into the test solution and taking the meter reading." It is the subject of this textbook to remind that this statement holds also true for E_H measurements. The procedure discussed in this chapter can also be obscured by a computer code like the microscopic processes at the interface of an electrode. The redox couple of As(III) – As(V) can be a very useful tool to estimate the redox status of a natural system, but in applying it everyone should be aware of the complex procedure and possible obstacles.

6.4 Acknowledgements

We would like to thank Mrs. U. Keller for careful reading of the manuscript by which she improves very much the English of the text. We would also like to thank two anonymous reviewers for their helpful comments on the manuscript.

6.5 References

AGGETT, J. & ASPELL, A.C. (1976): The determination of arsenic(III) and total arsenic by atomic-absorption spectroscopy. Analyst 101: 341-347.

AGGETT, J. & KADWANI, R. (1983): Anion-exchange method for speciation of arsenic and its application to some environmental analyses. Analyst 108: 1495-1499.

AGGETT, J. & KRIEGMAN, M.R. (1987): Preservation of arsenic(III) and arsenic(V) in samples of sediment interstitial water. Analyst 112: 153-157.

ANDERSON, R.K.; THOMPSON, M. & CULBARD, E. (1986): Selective reduction of arsenic species by continuous hydride generation. Part I: Reaction media. Analyst 111: 1143-1152.

ANDREAE, M.O. (1977): Determination of arsenic species in natural waters. Anal. Chem. 49: 820-823.

BALL, J.W. & NORDSTROM, D.K. (1991): WATEQ4F – User's manual with revised thermodynamic database and test cases for calculating speciation of major, trace and redox elements in natural waters. USGS Open-File Rep 90-129, 185 pp.

BLIEFERT, C. (1979): pH-Wert Berechnungen. VCH. Weinheim New York, 255 pp.

BÖKELEN, A. & NIESSNER, R. (1992): Removal of arsenic from mineral water. Vom Wasser 78: 355-362.

BRAMAN, R.S.; JOHNSON, D.L.; FOREBACK, C.C.; AMMONS, J.M. & BRICKER, J.L. (1977): Separation and determination of nanogram amounts of inorganic arsenic and methylarsenic compounds. Anal. Chem. 49: 621-625.

BROOKINS, D.G. (1988): Eh-pH diagrams for geochemistry. Springer. Berlin Heidelberg New York Tokio, 176 pp.

BROWN, D.S. & ALLISON, J.D. (1987): An equilibrium metal speciation model: users manual. Env. Res. Lab. Office of Res. and Dev. US EPA Rep. EPA/600/3-87/012, 103 pp.

CHAKRABORTI, D.; NICHOLS, R.L. & IRGOLIC, K.J. (1984): Determination of arsenite and arsenate by differential pulse polarography. Fresenius Z. Anal. Chem. 319: 248-251.

CHEAM, V. & AGEMIAN, H. (1980): Preservation of inorganic arsenic species at microgram levels in water samples. Analyst 105: 737-743.

CHERRY, J.A.; SHAIKH, A.U.; TALLMAN, D.E. & NICHOLSON, R.V. (1979): Arsenic species as an indicator of redox conditions in groundwater. J. Hydrol. 43: 373-392.

CUTTER, L.S.; CUTTER, G.A. & SAN DIEGO-MCGLONE, M.L.C. (1991): Simultaneous determination of inorganic arsenic and antimony species in natural waters using selective hydride generation with gas chromatography/photoionization detection. Anal. Chem. 63: 1138-1142.

DOVE, P.M. & RIMSTIDT, J.D. (1985): The solubility and stability of scorodite, $FeAsO_4 \cdot 2H_2O$. Am. Miner. 70: 838-844.

EBDON, L.; HILL, S.;WALTON, P. & WARD, R.W. (1988): Coupled chromatography - atomic spectrometry for arsenic speciation – a comparitive study. Analyst 113: 1159-1165.

GLAUBIG, R.A. & GOLDBERG, S. (1988): Determination of inorganic arsenic(III) and arsenic(III+V) using automated hydride-generation atomic-absorption spectrometry. Soil Sci. Soc. Am. J. 52: 536-537.

GRABINSKI, A.A. (1981): Determination of arsenic(III), arsenic(V), monomethylarsonate, and dimethylarsinate by ion-exchange chromatography with flame-less atomic absorption spectrometry detection. Anal. Chem. 53: 966-968.

HANSEN, S.H.; LARSEN, E.H.; PRITZL, G. & CORNETT, C. (1992): Separation of seven arsenic compounds by high-performance liquid chromatography with on-line detection by hydrogen-argon flame atomic absorption spectrometry and inductively coupled plasma mass spectrometry. J. Anal. At. Spectrom. 7: 629-634.

HASWELL, S.J.; O'NEILL, P. & BANCROFT, K.C.C. (1985): Arsenic speciation in soil-pore waters from mineralized and unmineralized areas of South-West England. Talanta 32: 69-72.

HUANG, Y.Q. & WAI, C.M. (1986): Extraction of arsenic from soil digests with dithiocarbamates for ICP-AES analysis. Commun. Soil Sci. Plant. Anal. 17: 125-133.

IVERSON, D.G.; ANDERSON, M.A.; HOLM, T.H. & STANFORTH, R.R. (1979): An evaluation of column chromatography and flameless atomic absorption spectrometry for arsenic speciation as applied to aquatic systems. Environ. Sci. Technol. 13: 1491-1494.

LÉONARD, A. (1991): Arsenic. In: MERIAN, E. (Ed.) Metals and their compounds in the environment. VCH. Weinheim, pp 751-774.

LINDBERG, R.D. & RUNNELLS, D.D. (1984): Ground water redox reactions: An analysis of equilibrium state applied to Eh measurements and geochemical modeling. Science 225: 925-927.

LINDER, P.W.; TORRINGTON, R.G. & WILLIAMS, D.R. (1984): Analysis using glass electrodes. Open University Press. Milton Keynes, 148 pp.

MASSCHELEYEN, P.H.; DELAUNE, R.D. & PATRICK JR., W.H. (1991): A hydride generation atomic absorption technique for arsine speciation. J. Environ. Qual. 20: 96-100.

NAUMOV, G.B.; RYZHENKO, B.N. & KHODAKOVSKY, I.L. (1974): Handbook of thermodynamic data (translated by SOLEIMANI, J.). USGS Wat. Res. Div. Rep. WRD 74-001, 328 pp.

ODANAKA, Y.; TSUCHIYA, N.; MATANO, O. & GOTO, S. (1985): Characterization of arsenic metabolites in rice plants treated with DSMA (disodium methanearsonate). J. Agric. Food Chem. 33: 757-763.

PARKHURST, D.L.; THORSTENSON, D.K. & PLUMMER, L.N. (1980): PHREEQE – A computer program for geochemical calculations. USGS Wat. Res. Invest. Rep. 80-96.

RAI, D.J.; ZACHARA, J.M.; SCHWAB, A.S.; SCHMIDT, R.; GIRVIN, D. & ROGERS, J. (1984): Chemical attenuation rates, coefficients, and constants in leachate migration. Vol I: A critical review. EPRI-EA-3356-Vol 1, 318 pp.

RÜDE, T.R. (1996): Beiträge zur Geochemie des Arsens. Karlsruher Geochem. H. 10: 206 pp.

SCHOELLER, H. (1962): Les eaux souterraines. Masson. Paris, 642 pp.

SPYCHER, N.F. & REED, M.H. (1989): Evolution of a Broadlands-type epithermal ore fluid along alternative P-T paths: Implications for the transport and deposition of base, precious, and volatile metals. Econ. Geol. 84: 328-359.

STUMMEYER, J.; HARAZIM, B. & WIPPERMANN, T. (1995): Arsen-Speziation mittels HPLC-Hydrid-AAS Kopplung. In: WELZ, B. (Ed.) Colloquium Analytische Atomspektrometrie. Bodenseewerk Perkin-Elmer. Überlingen. pp 425-428.

TALLMAN, D.E. & SHAIKH, A.U. (1980): Redox stability of inorganic arsenic(III) and arsenic(V) in aqueous solution. Anal. Chem. 52: 196-199.

VINK, B.W. (1996): Stability relations of antimony and arsenic compounds in the light of revised and extended Eh-pH diagrams. Chem. Geol. 130: 21-30.

WAGEMAN, D.D.; EVANS, W.H.; PARKER, V.B.; SCHUMM, R.H.; HALOW, I.; BAILEY, S.M.; CHURNEY, K.L. & NUTTALL, R.L. (1982): The NBS tables of chemical thermodynamic properties. J. Phys. Chem. Ref. Data 11 Suppl. 2. ACS AIP NBS. Washington, 392 pp.

WEIGERT, P. & SAPPL, A. (1983): Speciation of As(III) and As(V) in biological tissue. Fresenius Z. Anal. Chem. 316: 306-308.

WOHNLICH, S.; VOGELGSANG, A.; GLÄSSER, W. & DOHRMANN, H. (1997): Untersuchung hydrogeochemischer Prozesse am Beispiel der Braunkohletagebaukippe Zwenkau/Cospuden (Mitteldeutschland). In: Arbeitsgruppe des GBL-Gemeinschaftsvorhabens (Eds.) 3. GBL-Kolloquium. Stuttgart. pp 98-103.

Chapter 7

In Situ Long-Term-Measurement of Redox Potential in Redoximorphic Soils

S. Fiedler

7.1 Introduction

Soils can be defined as porous natural media. Depending on soil properties (e.g. continuity of pores) and climatic conditions, these pores can be either filled with air or water. The ratio between liquid and gaseous phase constrains the oxygen availability for dissimilation by micro-organisms and consequently, the redox conditions in soils.

However, irrespective of the medium, the quantitative use of measured redox potential (E_H-values) for thermodynamic calculations and modelling are generally limited. The required preconditions, such as

1. the reversibility of redox reactions at the surface of the electrode and
2. an equilibrium between redox partners and the electrode

are normally not completely given. Measured potentials generally represent the difference sum or integral of a magnitude of single reactions (mixed potential) in which several compounds are involved, which are crystallographically and thermodynamically not exactly defined (BRÜMMER, 1974).

The sequential reduction of redox pairs allows a definition of ranges (e.g. STUMM & BACCINI, 1978) of the most common redox reactions in the respective soils. One possibility of qualitatively assessing the nature of actual redox processes is to relate measured values to defined compounds. This information can be used to estimate eco-(toxico-)logical (e.g. nutrient availability, mobility of heavy metals) and pedogenetic processes and phenomena (colours, concretions and mottles).

Many investigators have measured E_H solely in soil suspensions or extracts. However, the validity of such measurements is limited due to the artificial conditions established. In natural soils, E_H varies strongly with time. Thus *in situ* and continuous measurements are preferable to trace E_H which is an important parameter to study transformation and translocation processes of redox sensitive (e.g. Fe, Mn) and associated elements (e.g. Mo, P, Co) in soils.

However, field measurements with permanently installed electrodes cause several problems. Besides general technical problems (e.g. the exchange current must be considerably stronger than the technically required current), especially soil specific features have to be taken into consideration:

1. A contamination of the electrode surface due to oxidation would lead to values influenced by the potential of the contaminant layer (e.g. PtO).
2. The pollution of the diaphragm of the reference electrode and/or the electrolyte solution would interrupt the galvanic contact between the soil and the electrode and would lead to a shift of the zero-point of the measuring system.

This chapter summarises the results of two research projects on transformation processes in redoximorphic soils related to redox conditions. To identify dominant redox conditions, we used continuously measured E_H at several field sites representing different degrees of redoximorphy.

7.2 Study Sites and Soils

The study sites are located in the moraine landscape of the cool-humid 'Allgäu' region (≈ 500 km^2), SW-Germany, between 550 m and 690 m a.s.l. Annual precipitation lies between 1200 and 1400 mm and annual mean air temperature is 6.5°C.

1. Long-term *redox measurements* (FIEDLER, 1997) were carried out (Siggen: 9°57'E, 47°42.5'N) at the wet margin of a pond at the lower end of a NNW exposed slope (9° inclination). For comparison we chose two sites in close proximity showing (i) pronounced differences in soil morphology, especially redoximorphism and (ii) marginal differences in altitudes (683.3 vs. 682.8 m a.s.l, distance = 5 m). According to the SOIL SURVEY STAFF (1997) the soils can be classified as Aeric Humaquept (AH) and Typic Humaquept (TH), respectively (see Table 7.1). The different intensity of redoximorphic features pointed to the presence of an intrapedonal E_H-gradient. Whereas in the Aeric Humaquept, Mn- and Fe-concretions revealed the dominance of oxidative conditions, the prevailing grey colour of horizons in the Typic Humaquept suggested essentially reductive conditions. E_H-values were measured to test this hypothesis.

2. The *emission of methane* (CH_4) was measured to test the hypothesis of increasing CH_4-emissions with increasing wetness and decreasing E_H. We chose three representative land units showing characteristic differences in wetness. In the following the drier soils are always given first:
   - Alluvial plain (Aichstetten, 10° 25'E 47° 52.2' N): Aeric Endoaquept (AE), Mollic Endoaquept (ME).
   - Wet colluvial margin of a pond (Artisberg, 9°E 51.5 47° 43' N): Typic Humaquept (TH), Fluvaquept Humaquept (FH).
   - Peaty depression (Wangen, 9° 50' E 47° 40.5' N): two Limnic Haplohemists one of which is drained (LHd) the other undrained (LHn).

Tab. 7.1: Soil Classification.

SOIL SURVEY STAFF (1997)	*BODENKUNDLICHE KARTIERANLEITUNG (1994)*
- Study 1 -	
Aeric Humaquept (AH	Humusreicher Oxigley
Ah1/Ah2/Bag/Bcg/Bg1/Bg2	Ah1/Ah2/AhGo/Gkso/Gro/Gor
Typic Humaquept (TH)	Humusreicher Naßgley
Ah1/Ah2/Bag/Bg1/Bg2/2Bg	Aa/Ah/AhGr/Gor/Gro/IIG(r)o
- Study 2 (Methane) -	
Aeric Endoaquept (AE)	Auengley
Mollic Endoaquept (ME)	Auennaßgley
Typic Humaquept (TH)	kolluvial überdeckter Anmoorgley
Fluvaquentic Humaquept (FH)	kolluvial überdeckter Moorgley
Limnic Haplohemist (drained site) (LHd)	entwässertes Niedermoor
Limnic Haplohemist (natural site) (LHn)	typisches Niedermoor

7.3 Methods

7.3.1 Redox Measurements

For the registration of E_H, the computer-controlled device developed by FIEDLER & FISCHER (1994) was used. Indicator electrodes (platinum) were self-made according to the instructions of MANN & STOLZY (1972). An Ag/AgCl-electrode was applied as a reference cell, the electrolyte solution (3.5 M KCl) was replaced every two months (KÖLLING, 1986). The indicator electrodes (n = 31, 2 -

3 per horizon) were installed horizontally into the profile walls. After installation, the profiles were carefully filled to approximately restore original conditions.

Measurements were carried out in hourly intervals over a period of 18 months. We corrected our initial redox measurement by the offset value of the hydrogen electrode standard. An adjustment to pH 7 was not carried out. Soil temperature was measured automatically in 5, 25, 40, 80 and 140 cm depths (thermocouples), precipitation and groundwater table were registered in hourly intervals (Delta-T-Devices LTD).

7.3.2 Methane Measurements

Fluxes of methane (CH_4) were measured weekly by a modified static chamber method (ROLSTON, 1986) for 2 years (in the following, results will be discussed only for the year 1997). At each site, two chambers (1 m^2 basal surface) were installed. Headspace samples were taken every 10 min over a period of 30 min after closure at 9 a.m., 12 and 3 p.m. Methane analyses were performed in a gas chromatograph (PE Autosystem XL) equipped with a flame ionisation detector. To measure the redox conditions, 20 electrodes were installed in each profile and at various depths. For measurements the electrodes were plugged to a portable pH/E_H-meter. Groundwater table as well as E_H was monitored weekly.

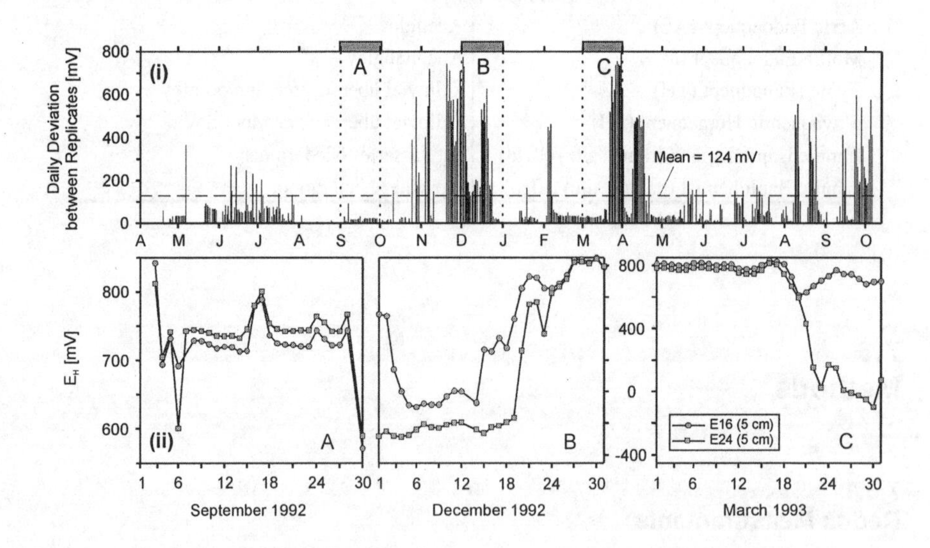

Fig. 7.1: (i) Daily difference of E_H-values between two electrodes 10 mm apart and (ii) the E_H-values (daily mean) of replicates in a depth of 5 cm (AH, Ah1).

7.3.3 Spatial and Temporal Variability of E_H

The heterogeneous distribution of reactive substances (micro-habitats, hot spots, microsites) was often attributed to insufficient reproducibility of parallel measurements (e.g. PARKER et al., 1985). According to FEAGLEY (quoted in BLANCHAR & MARSHALL, 1981), deviations of 50 to 60 mV, are still acceptable. BLUME (1968) estimated the temperature and pH dependent error amounting up to ± 100 mV, while MCKEAGUE (1965) observed deviations of up to 300 mV.

In addition to the correlation coefficients, which give a measure regarding the similarity of values, the differences of measured values between two electrodes may be used to identity different E_H-levels. Our results demonstrate that the differences between two electrodes may be highly variable over time. For example, differences between the replicates of daily ΔE_H-values in Ah1 (AH) ranged between 0 and 800 mV, with a mean deviation of 124 mV. Three characteristics for two replicates (both installed at 50 mm depth with a horizontal distance of 10 mm) were observed (Figure 7.1):

1. similar trend and level (ΔE_{Hmin} = 10 mV; ΔE_{Hmax} = 100 mV),
2. similar trend; but different level (ΔE_{Hmax} = 760 mV),
3. both trend and level are different (ΔE_{Hmax} = 800 mV).

High groundwater tables, and periods of snow cover (November - December 1992) or melt (mid-March - mid-April 1993) were always accompanied by the highest differences between replicates. This was also the case, when the groundwater table was fluctuating within one horizon (see Figure 7.7), separating it into a saturated and an unsaturated zone. When the water level dropped below the meas-

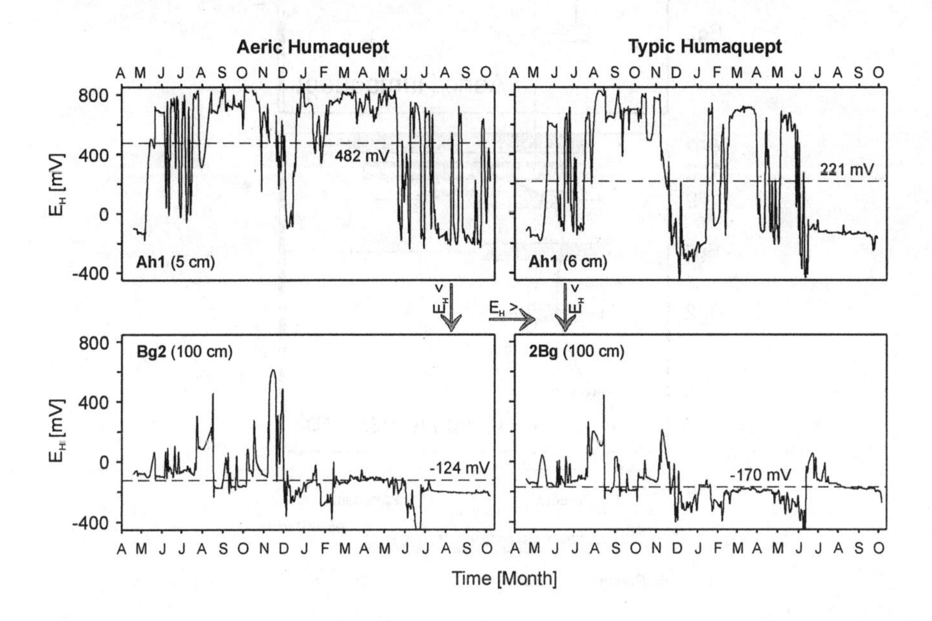

Fig. 7.2: Mean daily E_H-values from April 1992 to October 1993 for surface and subsoil horizons of two soils. The median of the E_H is shown as a dashed line.

uring zone, an erratic increase of up to 900 mV was observed with one day delay, similar to that observed by MANSFELDT (1994).

These results shows, that E_H may fluctuate within small horizontal distances. For this reason, the generally applied practise of several measurements at the same depth seems unsuitable to obtain a meaningful mean value unless the spatial resolution would be extremely enhanced. However, they are suitable to reflect the specific ranges of E_H-values within a horizon. Consequently, E_H-measurements must be considered as local measurements, only representative for the surrounding soil matrix at a scale approximately of 1 mm^3.

Figure 7.2 gives an example of an E_H-curve (here presented as daily means) in a top- and a subsoil horizon of the two soil profiles, respectively. The two bioactive active surface horizons exhibit the largest E_H-variation. We observe (i) a decrease of E_H-levels and E_H-amplitudes with increasing depth, and (ii) an E_H-gradient between both soils. The time period with reducing conditions is larger for the Typic than for the Aeric Humaquept.

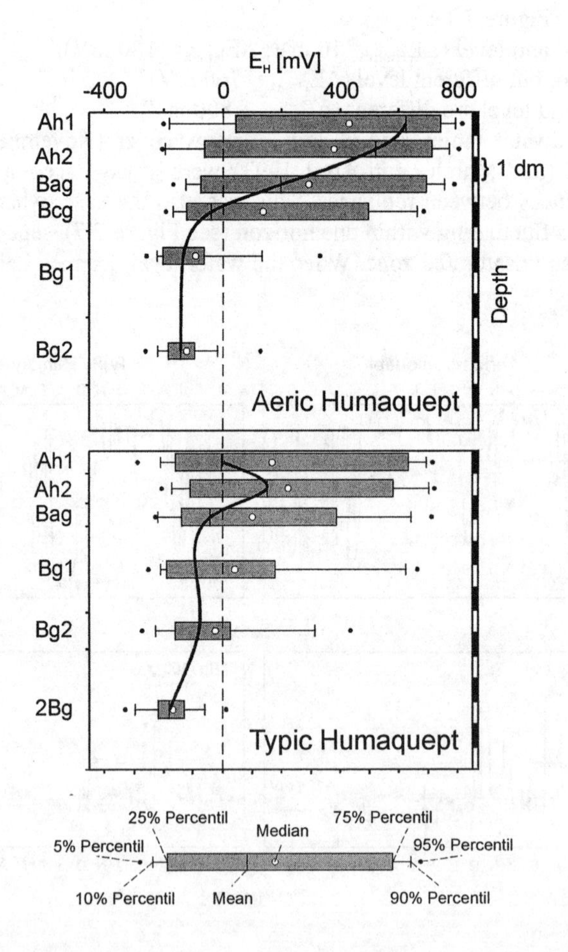

Fig. 7.3: Descriptive statistics of E_H-values from April 1992 to October 1993 for two soils.

The medians (Figure 7.2) suggest that methane is the predominant redox product for the Typic Humaquept. This is supported by KLEBER (1997), who measured methane emissions of up to 30 mg CH_4 $m^{-2} \cdot d^{-1}$ at the same place. A further consequence of different redox conditions are relative losses and gains of pedogenic Mn-oxide when compared to the corresponding contents of the parent material. By using the ratio-method (SOMMER & STAHR, 1996), we calculated a relative loss of 46 % manganese (Mn) in the Typic Humaquept, whereas the Aeric Humaquept accumulated 5 % Mn. With exception of the 2Bg horizon, the E_H-gradient fits very well to soil morphology (soil colours, mottles). In the 2Bg horizon, the prevailing features cannot be explained by actual E_H-conditions due to oxidation.

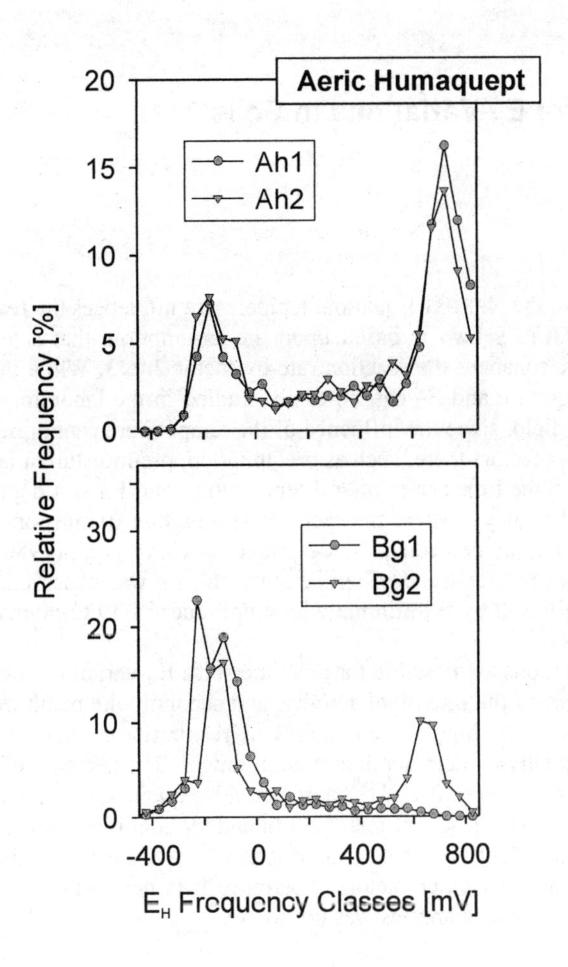

Fig. 7.4: Relative frequency of E_H-values from April 1992 to October 1993 for A- and B-horizons (Aeric Humaquept).

So we concluded that this horizon is not in equilibrium with the present conditions, probably due to the recent rise of the pond water level. This means that E_H-measurements may also be applied as an indicator of relictic genetic features.

Figure 7.3 shows the frequency and distribution of E_H-values in the form of box plots. The deviation between the median and the mean value (e.g. for Ah1 of the Aeric Humaquept this is 195 mV) suggests an asymmetric distribution. When the values of the biologically most active A-horizons are assigned to frequency classes (50 mV), a bimodal distribution is obtained, which can be considered as proof of asymmetry (Figure 7.4). In these horizons, the most frequent classes depend on the season (not presented). For B-horizons which are water saturated over the entire period, the E_H does not exhibit any seasonal response.

7.4 Reasons for E_H-Variations in Soils

7.4.1 Temperature

According to the NERNST Equation, temperature influences the reaction equilibrium. VAN'T HOFF'S Law is based upon the assumption, that a temperature increase of 10°K enhances the reaction rate by factor 2 to 3. While the relationship between temperature and E_H can be easily studied in the laboratory, this is more difficult in the field, since the influence of the temperature cannot be readily separated from other factors there, such as precipitation, pre-moisture, plant cover, etc.

In Figure 7.5 the time series of soil temperature and E_H at a depth of 5 cm are presented for five days, where the factors precipitation (0 mm) and groundwater table (90 cm) may be neglected. As described by CLAY et al. (1980) and FARELL et al. (1991), an inverse trend of temperature and E_H was observed. Temperature maxima are followed by E_H-minima with amplitudes of 30 to 50 mV and a phase shift of 6 hours.

Two explanations are possible for this time shift: E_H-variations are temperature-induced changes of the microbial activity, and secondly the result of root respiration (which is much higher then microbial respiration). This leads to an E_H-minimum especially in 5 cm depth around midnight. The decrease of the nocturnal root respiration between 0 and 12 a.m. would explain the gradual E_H-increase within this time. As a result of changing boundary conditions, such as rise of the groundwater table (not presented), the relation between temperature and E_H may be often concealed by other factors. Therefore it is not surprising that a simple relation between these quantities was not found.

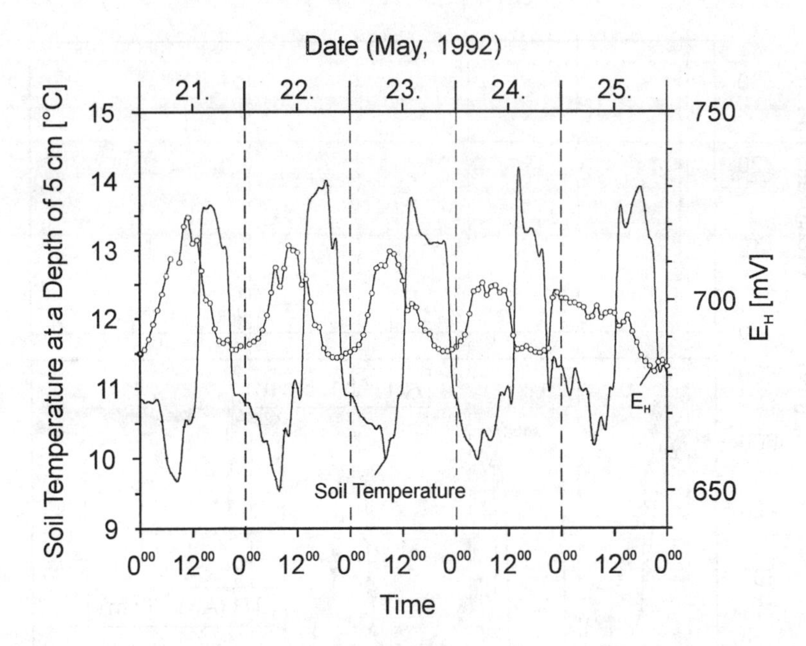

Fig. 7.5: Relationship between temperature (at a depth of 5 cm) and E_H (AH: Ah1) during the period from May 21 to May 25, 1992.

7.4.2 Water Regime

7.4.2.1 Precipitation

Up to now, the influence of precipitation on soil E_H has not been completely understood. HAAVIST (1974) detected a decrease in E_H, independent of pH, following heavy precipitation (however without reporting rain intensities) of 48 mV. In contrast, WIECHMANN (1978) found an E_H-increase after irrigation which he attributed to the high O_2-contents of the irrigation water.

E_H-variation of the uppermost horizons of the two soils after different precipitation events are presented in Figure 7.6. These events, lead to initial short-term increases of the E_H, which were obviously due to the O_2-contents in rainwater (34 Vol.% at 1013.24 mbar and 10°C). These phases are followed by pronounced decrease of E_H by 100 to 200 mV/h within short time. After the rainfall stopped, E_H rapidly increased again by 500 mV. This may be attributed to rapid drying of the soil surface as well as to good aeration. In depths >30 cm, precipitation influenced the potentials only indirectly by rising the water table.

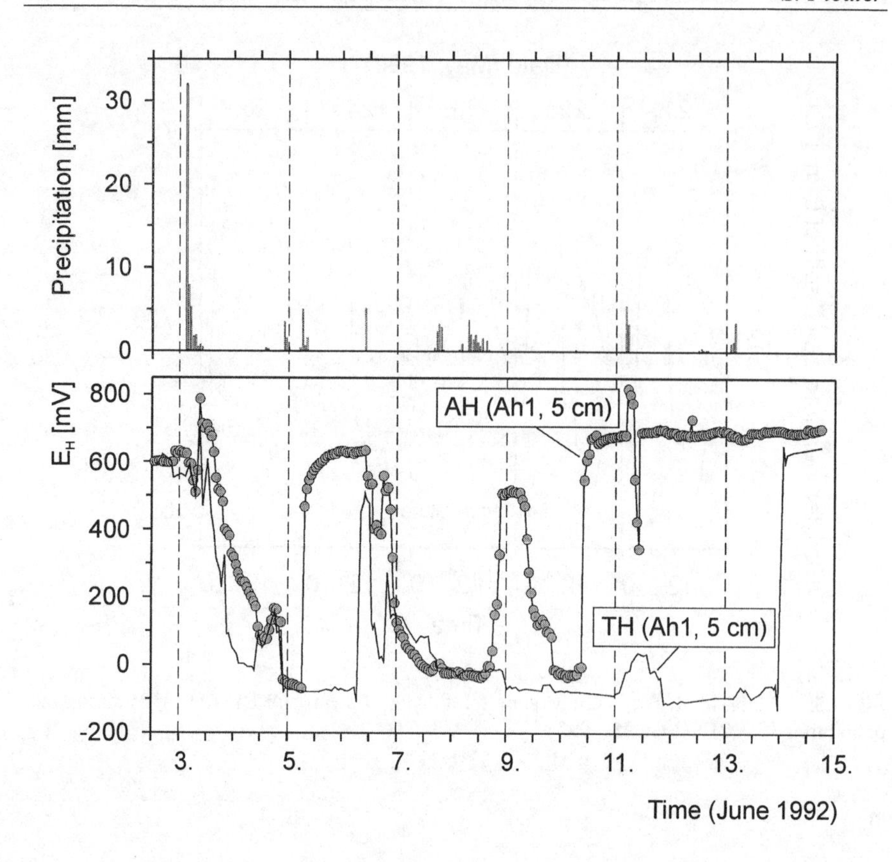

Fig. 7.6: Relationship between precipitation and E_H for the A-horizon of the Aeric Humaquept (AH) and Typic Humaquept (TH) for the period from June 3 to June 15, 1992.

As compared to the Aeric Humaquept, the Typic Humaquept (i) responds faster to precipitation and (ii) potentials remained in the negative range for longer periods. This can be attributed to a different position (Typic Humaquept in a depression site, additional lateral supply of water), as well as to soil physical properties (Typic Humaquept larger pore volume with less air capacity, hence lower diffusive transport of O_2).

A close relationship between precipitation and E_H was supposed, but the correlation with values between -0.1 and -0.3 did not support this hypothesis, even when a possible vertical retardation was considered.

This might be attributed to changing boundary conditions, which are difficult to parameterise. The formation of inhomogeneous watering fronts mainly depends on pre-moisture conditions (McCORD et al., 1991). In comparison, other factors, such as rainfall intensity, evaporation, root growth (BEVEN, 1991) and soil physical properties, are of minor importance. The results suggest that the infiltration rate and its influence on the temporal and spatial variability of E_H cannot be interpreted without more detailed information on the pore system, which is identical with the small scale distribution of water.

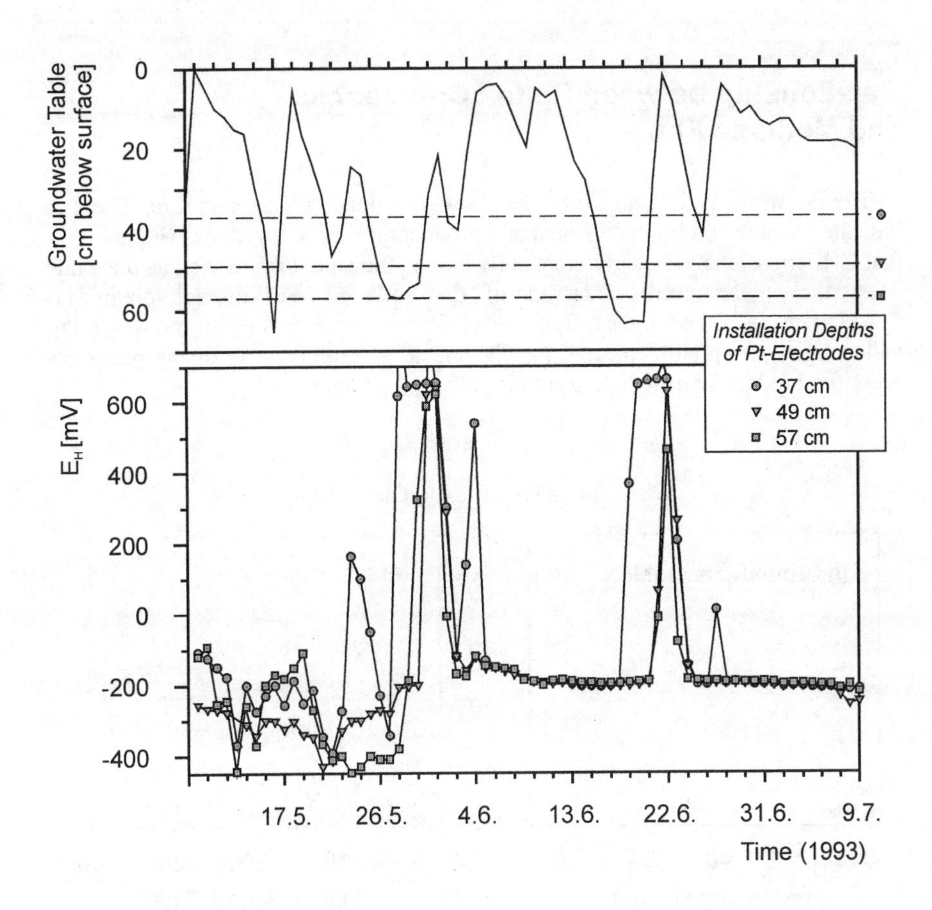

Fig. 7.7: Relationship between groundwater table and E_H (AH, Bg2) for the period from May 17 to July 27, 1993.

7.4.2.2 Groundwater

The inverse trend of groundwater level and E_H are evidenced by numerous observations (e.g. FAULKNER et al., 1989). The alteration of E_H following water saturation is a direct consequence of O_2-consuming microbial processes (WHISLER et al., 1974). The groundwater table has only an indirect influence on E_H via the air regime (O_2-displacement, O_2-diffusion). Among all parameters monitored, the groundwater table shows the closest relationship to E_H (r^2 = -0.42). The trend towards negative correlation between both variables was decreased with depth.

7.5
Relationship between Redox Conditions and Methane Flux

The E_H, which is determined by the above-mentioned parameters should help to indicate suitable conditions for methane production. The study shows (Figure 7.8), that CH_4-emissions varied systematically among the land units as well as the study sites: the lowest were observed in the soils of the alluvial plains (1, 2.9 g $CH_4 \cdot m^{-2} \cdot a^{-1}$). The highest fluxes were determined from soils of the wet colluvial margin of a pond (50.9, 90.9 g CH_4 $m^{-2} a^{-1}$). Haplohemists of the peaty depressions held an intermediate position (2.8, 14.4 g CH_4 $m^{-2} a^{-1}$).

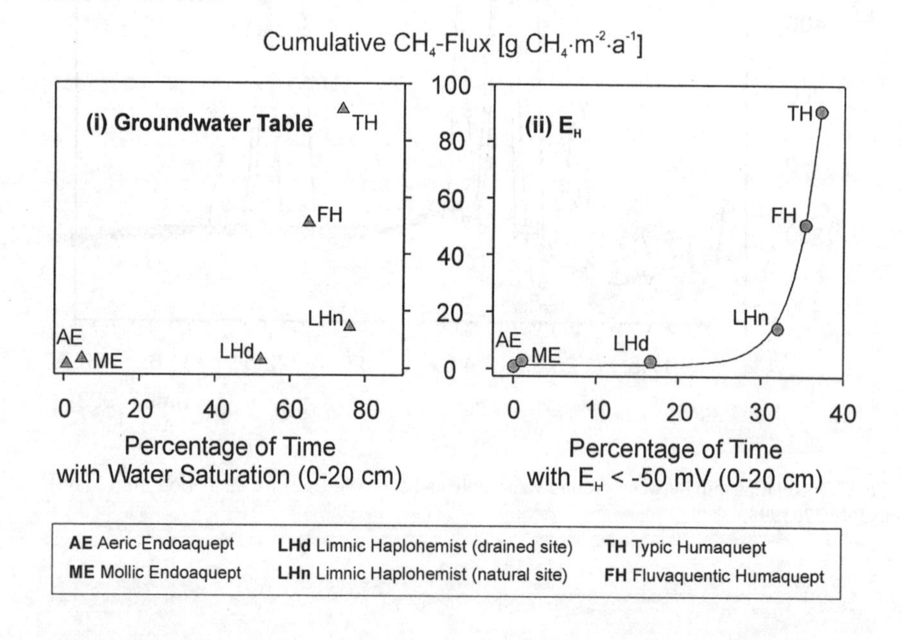

Fig. 7.8: Relationship between (i) groundwater table (median) and cumulative methane flux, (ii) time percentage of $E_H < -50$ mV in the topsoil and cumulative methane flux.

There is a clear increase in groundwater table from the Aeric to Mollic Endoaquepts to the Humaquepts and Haplohemists, respectively (Figure 7.8), whereas the soils of the colluvial margin (groundwater level 10 cm below surface) and the peaty depression (groundwater level 7 cm below surface) exhibit similar wetness.

Their methane emissions cannot be explained by the groundwater table alone. However a consistent relation is obtained, when reductive conditions in the layer above the groundwater level are considered. In Figure 7.8 the percentage of the time with E_H-values below a threshold value of –50 mV (above this level it is assumed that no methane oxidation occurs, NEUE & ROGER, 1994) is plotted against the cumulative CH_4-emissions. Both variables are highly correlated showing a non-linear relationship: The longer the period of E_H-values lower than -50 mV, the higher the fluxes.

Consequently, methane emission is determined by the reducing conditions and the thickness of the zone above the groundwater level: When E_H-values indicate oxidising conditions and this zone is thick, methane can be oxidised while diffusing through this zone. In this case, there might be no detectable methane emission detectable despite of its considerable production in deeper layers.

7.6 Conclusion

Our study confirms that the characterisation of the dominant redox processes may be derived from *in situ* measurements of E_H. These measurements lead to interpretable results with regard to the role of abiotic factors controlling E_H. The following sequence could be elaborated with regard to the influence of single factors on E_H:

groundwater table >> temperature > precipitation.

This study shows a high spatial and temporal variability of E_H. Though the E_H may fluctuate within small horizontal distances of 1 cm between 0 to 800 mV, the E_H-values are suitable to reflect the specific ranges within a horizon.

Measurements of E_H can provide useful information related to processes of soil development as well as translocation. For example, element loss in case of Mn shows a strong relationship to the time period of reducing conditions. Long-term reaction of the pedogenic oxides seems to be in agreement with the actual redox conditions. Measurements of E_H can be considered an important instrument for the classification of redoximorphic soils. Low E_H-levels were accomplished by increasing the methane flux. We found a strong association between redoximorphic features and redox conditions.

7.7 References

BEVEN, K. (1991): Modelling preferential flow: An uncertain future? In: GISH, T.J. & SHIMOHAMMADI, A. (Eds.): Preferential Flow Proceedings of the National Symposium, 16/17 December 1991 Chicago, Illinois, p. 1-11.

BLANCHAR, R.W. & MARSHALL, C.E.(1981): Eh and measurement in Menfro and Mexico soils. ASA Spec. Publ. 40: 103-128.

BLUME, H.P. (1968): Stauwasserböden. Arbeiten der Univ. Hohenheim 42, Ulmer Verl., Stuttgart, 242 p.

BODENKUNDLICHE KARTIERANLEITUNG (1994): Ad-hoc-Arbeitsgruppe Boden, 4.Aufl., Hannover, 392 p.

BRÜMMER, G. (1974): Redoxpotentiale und Redoxprozesse von Mangan-, Eisen- und Schwefelverbindungen in hydromorphen Böden und Sedimenten. Geoderma, 12: 207-222.

CLAY, D.E.; CLAPP, C.E.; MOLINA, J.A.E. & LINDEN, D.R. (1980): Soil tillage impact on the diurnal redox-potential cycle. Soil Sci. Soc. Am. J. 54: 516-521.

FARELL, R.E.; SWERHONE, G.D.W. & VAN KESSEL, C. (1991): Construction and evalua-tion of a reference electrode assembly for use in monitoring in situ soil redox potentials. Commun. in Soil Sci. Plant Anal. 22: 1059-1068.

FAULKNER, S.P.; PATRICK, W.H. & GAMBRELL, R.P. (1989): Field techniques for measuring wetland soil parameters. Soil Sci. Soc. Am. J. 53: 883-890.

FIEDLER, S. & FISCHER, W.R. (1994): Automatische Meßanlage zur Erfassung kontinuierlicher Langzeitmessungen von Redoxpotentialen in Böden. Z. Pflanzenernähr. Bodenkd. 157: 305-308.

FIEDLER, S. (1997): In-situ-Langzeitmessungen des Redoxpotentials in hydro-morphen Böden einer Endmoränenlandschaft im württembergischen Alpenvorland. Hohenheimer Bodenkdl. Hefte 42, 135 p.

HAAVIST, V.F. (1974): Effects of a heavy rainfall on redox potential and acidity of a waterlogged peat. Can. J. Soil Sci. 54: 133-135.

KLEBER, M. (1997): Cabon exchange in humid grassland soils. Hohenheimer Bodenkdl. Hefte 41, 264 p.

KÖLLING, M. (1986): Vergleich verschiedener Methoden zur Bestimmung des Redoxpotentials natürlicher Wässer. Meyniana 38: 1-19.

MANN, D.L. & STOLZY, L.H. (1972): An improved construction method for platinum microelektrodes. Soil Sci. Soc. Amer. Proc. 36: 853-854.

MANSFELDT, T. (1994): Schwefeldynamik von Böden des Dittmarscher Speicherkoogs und der Bornhöveder Seenkette in Schleswig-Holstein. Schriftenreihe Inst. Pflanzenernähr. Bodenkd. Univ. Kiel 28, 155 p.

MCCORD, J.T.; STEPHENS, D.B. & WILSON, J. (1991): Towards validating state dependent macroscopic anisotropy in unsaturated media: Field experiments and modelling considera-tion. J. of Containment Hydrology 16: 145-175.

MCKEAGUE, J.A. (1965): Relationship of water table and Eh to properties of three clay soils in the Ottawa Valley. Can. J. Soil Sci. 45: 49-62.

NEUE, H.-U. & ROGER, P.A. (1994): Rice agriculture: Factors Controlling Emissions: 254-298. In: KHALIL, M. (Ed.): Atmospheric Methane: Sources, Sinks, and Role in Global Chance. Springer-Verlag, Bln.

PARKER, W.B.; FAULKNER, S.P. & PATRICK, W.H. (1985): Soil wetness and aeration in selected soil with aquatic moisture regimes in the Mississippi and Pearl River Deltas. In: Wetland soils: Characterisation, classification and utilisation. Intern. Rice Res. Inst. Los Banos: pp. 91-107.

ROLSTON,D.E. (1986): Gas flux. p. 1103-1119 In: Klute, A. (ed.): Methods of Soil Analysis, Part I, 2nd Edition, Madison, Wisconsin.

SOIL SURVEY STAFF (1997): Keys of taxonomy. 8th edition, 644 p.

SOMMER, M. & STAHR, K. (1996): The use of element: clay-ratios assessing gains and losses of iron, manganerse and phosphorus in soils on a landscape scale. Geoderma 71: 173-200.

STUMM, W. & BACCINI, P. (1978): Man-made perturbation of lakes. In: LERMAN, A. (Ed.): Lakes Chemistry, Geology, Physics. Springer Verl., Berlin, Heidelberg, New York, p. 91-126.

WHISLER, F.D.; LANCE, J.C. & LINEBARGER, R.S. (1974): Redox potentials in soil columns intermittently flooded with sewage water. J. Environ. Quality 3: 68-74.

WIECHMANN, H. (1978): Stoffverlagerung in Podsolen. Ulmer Verl., Stuttgart. 134 p.

Chapter 8

Redox Measurements as a Qualitative Indicator of Spatial and Temporal Variability of Redox State in a Sandy Forest Soil

A. Teichert, J. Böttcher & W.H.M. Duijnisveld

8.1 Introduction

In a groundwater recharge area in Northwest Germany, measurements have been carried out for more than two decades, in order to quantify the solute transport and solute transformations in soils and aquifer. One objective of these studies is to predict the quality development of groundwater recharge. In order to achieve this, among other things, landuse-specific solute transport in sandy soils and solute leaching into groundwater must be known. That is why spatial and temporal variability of water and solute fluxes from soils into the groundwater is studied intensively.

Around 40% of the recharge area of the waterworks at our research site is covered by coniferous forest. Below coniferous forests, low pH-values and high Al^{3+} and SO_4^{2-} concentrations are typical for groundwater recharge, so acidification has reached the groundwater (STREBEL et al., 1993; FRANKEN et al., 1997). But the acid forest soils and the unsaturated zone underneath are not only a transport medium but also a temporally variable and spatially heterogeneous reservoir for acidity and metals, especially H^+, Al^{3+}, Fe^{3+} and SO_4^{2-} (PUHLMANN et al., 1997; BÖTTCHER et al., 1997).

Beside this storage function, on sites with temporally high groundwater table, reduction and oxidation reactions also may occur in that depth zone of the soil in which the degree of saturation changes periodically. Those redox processes can change the chemical conditions of the soil permanently. As outlined by VAN BREEMEN (1987) redox processes can cause strong acidification or alkalisation in soils and aquatic environments. Oxidised components of redox pairs are usually more acidic than their reduced counterparts (e.g. H_2SO_4 vs. H_2SO_3; $pK_a(H_2SO_4) \approx -3$; $pK_a(H_2SO_3) = 1.96$). As a result alkalinity and pH of soil and water tend to increase upon chemical reduction and to decrease upon oxidation. On the other hand, changes in soil capacity factors like the Acid Neutralising Capacity (ANC) and Base Neutralising Capacity (BNC) may become permanent, if one of the redox processes leads to the formation of pairs of components of different mobility. Removal of one of the components leads to a system that has become permanently acidified or alkalinised. An example is the formation of acid sulfate soils in marshlands. Under waterlogged conditions sulfate and sedimentary iron are reduced to FeS_2 that can be considered as immobile potential acidity, and HCO_3^- that can be considered as mobile alkalinity. This mobile alkalinity may disappear (e.g. by leaching), whereas oxidation of FeS_2 after soil aeration leads to strong soil acidification (VAN BREEMEN, 1987). In acidic forest soils sulfate has to be considered in particular. As the predominant mobile anion in the soil solution of aerated forest soils, it promotes leaching of M_b (= base) cations. In anaerobic soils, reduction of sulfate may be important for the neutralisation of acids. Whenever soils are under the influence of groundwater, redox reactions can be expected. Therefore, redox measurements should be included in the examinations of interactions between the acidification of soils and water (KRUG, 1991).

One practicable way of indicating the redox state in soils is to measure the redox potential (E_H). In general, this is carried out using platinum electrodes. However, due to several problems, the value of such measurements in natural systems is highly limited (e.g. SPOSITO, 1989; PEIFFER et al., 1992; STUMM & MORGAN, 1996; SIGG & STUMM, 1996). In soils, E_H-measurements indicate the redox state only qualitatively (SPOSITO, 1989). However, clues to possible redox processes and associated proton release can be deduced. After specific additional measurements (such as solute concentrations, pH-values etc.) and geochemical modelling, these processes can be further quantified. Due to temporal changes of the groundwater table, fluctuating redox states in groundwater soils are to be expected. Redox potential measurements have to be carried out over a sufficiently long period of time, in order to be able to record periods of different redox states.

Beside the temporal dynamics however, a spatial heterogeneity of solute concentrations and solute fluxes in forest soils can also be expected. BÖTTCHER & STREBEL (1988b) found that sulfate concentrations and pH-values of groundwater recharge showed considerable spatial variations. Approximately half of the variations could be traced back to systematic spatial fluctuations of the solute input with the canopy throughfall (BÖTTCHER & STREBEL, 1990). The other half was caused by a soil-internal variability, for which not only the heterogeneity of the chemical soil properties (TEICHERT et al., 1997) but also of the physical soil properties (DEURER et al., 1997) have to be considered. The spatial variability of soil properties, particularly of material stocks in soils (e.g. organic carbon, TEICHERT et al., 1997), and the influence of fluctuating groundwater table could cause spatially variable redox potentials and conditions. These could, via proton releases, take an influence on the spatial variability of pH-values and the acidity of groundwater recharge.

The intention of our investigations is to measure on site the temporal and spatial heterogeneity of redox potentials. It might indicate the redox conditions in a groundwater-influenced acid sandy soil underneath a coniferous forest. From these data, conclusions are to be drawn in respect to the temporal and spatial differentiation of oxidising or reducing conditions in soils along a 10 m transect, and its potential influence on transformation processes. Particularly proton consumption and production is to be examined.

8.2 Material and Methods

8.2.1 Investigation Site

The investigation site is located in the *Fuhrberger Feld* about 30 km north-east of Hannover, in the plains of late Pleistocene fluviatile sands of Northwest Germany. It is a typical water recharge area in the north of Germany. Average annual rainfall is about 680 mm, average annual temperature is 8.9°C. At the investigation site the sandy soil is a Gleyic Podzol (FAO-System), formed as fine to me-

Tab. 8.1: Statistical parameters of soil chemical properties in a sandy forest soil (Gleyic Podzol).

	C_{org}		*BNC*		*CEC*		*pH(H₂O)*		*pH(CaCl₂)*	
Depth[m]	X	CV	X	CV	X	CV	X	CV	X	CV
0.15	4.25	24	95.7	26	43.1	25	3.68	3	2.77	4
0.30	0.71	85	19.3	103	10.7	76	4.19	4	3.12	6
0.45	3.60	29	132.0	24	56.1	25	3.80	3	3.14	5
0.60	1.29	41	52.0	36	16.1	41	4.35	2	3.94	5
0.75	0.60	52	19.1	44	7.8	36	4.53	2	4.15	3
0.90	0.29	55	11.4	35	5.0	35	4.68	2	4.25	4
1.05	0.16	35	5.9	41	3.6	46	4.81	3	4.34	3
1.20	0.12	42	7.0	47	3.8	53	4.76	2	4.39	4
1.35	0.14	56	8.0	65	5.6	81	4.73	3	4.31	5
1.50	0.12	20	5.2	27	4.0	20	4.67	4	4.22	2

X: Arithmetic mean; CV: Variation coefficient in %; C_{org}: organic carbon in %; BNC: Base Neutralising Capacity ($mmol_c$/kg); CEC: Cation Exchange Capacity ($mmol_c$/kg)

dium sands. Some important soil characteristics are given in Table 8.1. In the course of one year, the groundwater table may fluctuate between 0.7 and 2.0 m beneath surface.

8.2.2 Transect Measurements

In early 1997 a transect with a length of 10 m was set up along a row of 70-year-old pine trees. At the depth of main groundwater table fluctuations (0.8 m, 1.1 m and 1.4 m) 57 platinum electrodes for redox potential measuring were installed. Furthermore, antimony electrodes for pH-measuring and porous suction cups for sampling the soil solution were installed at 4 different depths (0.2 m, 0.5 m, 0.8 m and 1.1 m). In December 1997, additional suction cups and Sb-electrodes were built in at 1.4 m depth. The horizontal distance between transect positions of the probes is 0.5 m. The horizontal and vertical distribution of the different probes is shown in Figure 8.1. To minimise possible water flow along the tubes of the probes, redox- and pH-electrodes as well as suction cups were installed at an angle of 45°. Beneath the transect, groundwater table fluctuations are continuously measured. Canopy throughfall is sampled weakly along the transect at the 19 positions by precipitation collectors (bottles of polypropylene with funnels and a volume of 2 litres).

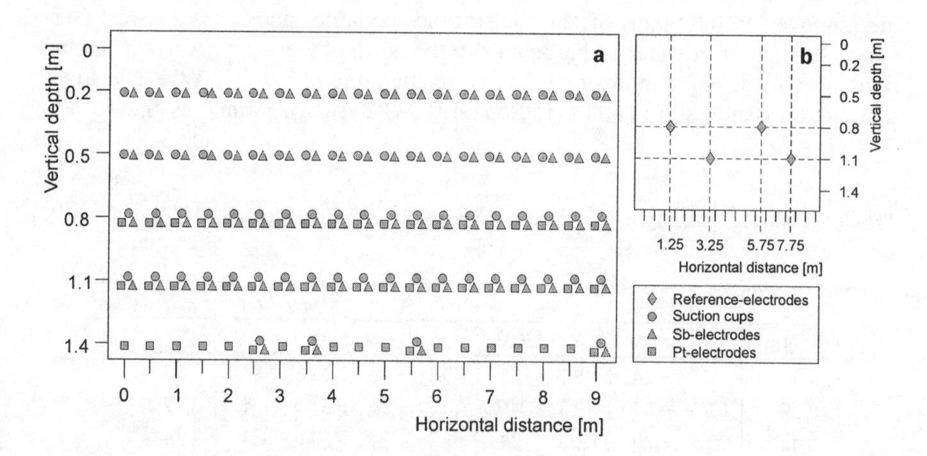

Fig. 8.1: Horizontal and vertical positions of the different probes: (a) at the transect; (b) approximately 0.7 m in front of the transect (reference electrodes).

8.2.3 Electrodes

Platinum electrodes (Pt-electrodes) are used to measure redox potential. The electrodes were constructed following the description of MUELLER et al. (1985) and PFISTERER & GRIBBOHM (1989). Sintering the platinum together with the wires to a ceramic jacketing prevents an intrusion of water. The active surface of the Pt-tip amounts to 30 – 35 mm^2. Tests concerning the functioning of the platinum electrodes in a redox buffer showed deviations <5 mV from the given potential of the buffer solution.

pH-values are measured with antimony electrodes (Sb-electrodes). In contrast to the microelectrodes for pH-measurement as described by MARTH & LOFTFIELD (1991), industrially produced antimony splinters were used. Each antimony electrode was calibrated in 'soil pastes' of soil material from the investigation site (MARTH & LOFTFIELD, 1991).

Four Ag/AgCl-reference electrodes (Ingold, InLab 301 Reference) were installed approximately 0.7 m in front of the transect. In order to rule out malfunctioning of the reference electrodes by low temperatures in winter months, installation depths of 0.8 m and 1.1 m were chosen (Figure 8.1). Preliminary examinations at the site revealed, that the depth of the reference electrodes (0.2 to 1.4 m) does not affect the measured potentials. The Pt-, Sb- and reference probes are connected by KCl-Agar salt bridges (VENEMANN & PICKERING, 1983). Malfunctioning of the reference system can be caused by the compression of the salt bridge (MANSFELDT, 1993). Furthermore, fluctuations of the measured potentials can be attributed to the usage of aged electrolyte solution (see Chapter 4). Because of that, the long-term stability of the reference potential is limited in time. Therefore, salt bridges are changed periodically every 3 months (see MANSFELDT, 1993). On this occasion the functioning of the reference system is also controlled, and fresh electrolyte solution is refilled, if required.

E_H-, Sb- and reference-potentials were automatically recorded by a self-constructed data logger. The potentials were measured and stored every hour. Because of the technical design the single potential is measured primarily against a metal stake (ground stake). In this way potential fluctuations at the reference electrodes are considered. E_H- and Sb-potentials are calculated by equalising measured potentials against each reference probe. Finally, the stored value is the estimated arithmetic mean of the considered potential differences.

In the following, daily mean values of pH and E_H, instead of hourly values are given for the reasons of clarity.

8.2.4 E_H- and pε-Values

Redox potentials are given in this text as E_H-values in [mV]. Because of its similarity to the pH the redox potential is expressed in some textbooks by the negative logarithm of the free-electron activity, the pε-value ($p\varepsilon \equiv -\log(e^-)$; SPOSITO, 1989). To compare the measured redox potentials with values given in literature, and to classify the redox state of the soil, pε values were calculated from E_H-values by (SPOSITO, 1989):

$$p\varepsilon = \frac{E_H}{59} \tag{8.1}$$

8.2.5 Statistical Data Analysis

The time series of redox potentials were analysed by spectral variance analysis, often simply called spectral analysis. Spectral analysis as a part of time series analysis can be used to detect periodic structures of spatial or temporal data series (JENKINS & WATTS, 1968; SCHÖNWIESE, 1985; BÖTTCHER & STREBEL, 1988a). The variance of a series of measurements is split up into individual frequency-dependent contributions and is graphically represented as a power spectrum. Peaks in the power spectrum indicate frequencies for periods p = 1/f [day] of pronounced periodicity in the data series (BÖTTCHER & STREBEL, 1988a).

The calculation of the power spectrum is based on the FOURIER transformation of the auto-correlation function. The spectral power for a given frequency is estimated as:

$$R(f) = 2\left\{1 + \left[\sum_{h=1}^{L-1} r(h)\, w(h) \cos(2\pi f h)\right]\right\} \tag{8.2}$$

where:

R(f) = spectral power [dimensionless]
L = maximum lag h (truncation point) [day]
h = time lag between pairs of measurements [day]
r(h) = auto-correlation function [dimensionless]
w(h) = weighting function, called lag window [dimensionless]
f = frequency [day^{-1}].

We used the TUKEY lag window as weighting function in (8.2) (BÖTTCHER & STREBEL 1988a):

$$\begin{aligned} w(h) &= \left\{\frac{1}{2}\left[1 + \cos\left(\frac{\pi h}{L}\right)\right]\right\}, & 0 \le h \le L \\ w(h) &= 0, & h > L \end{aligned} \tag{8.3}$$

For selective representation of dominant periodicity in data series the method called FOURIER transform smoothing was applied. This method eliminates disturbing scatter in data series by a smoothing function. BÖTTCHER & STREBEL (1988a) give a detailed description of the calculations.

8.3 Results and Discussion

The data analysis presented here is based on data received in the measuring period from March 1997 until September 1997. The gap in the data series at the beginning of April is due to a short-term break-down of the measuring devices.

8.3.1 E_H and pH at a Depth of 0.8 m and of 1.1 m

The redox potential of the individual electrodes at a depth of 0.8 m and of 1.1 m show relatively few fluctuations during the investigation period. The absolute values (see Figure 8.2, E_H depicted in relation to pH) range between 600 mV and 850 mV. Due to the fact that an increasing spatial and temporal variability of redox state is not deducible from the potential courses, there is no need for a presentation of the redox potential series of individual electrodes. The data series of the redox potentials (Figure 8.3) averaged over all electrodes of 0.8 m and 1.1 m depth level clearly reflect the few changes of redox status in the investigation period.

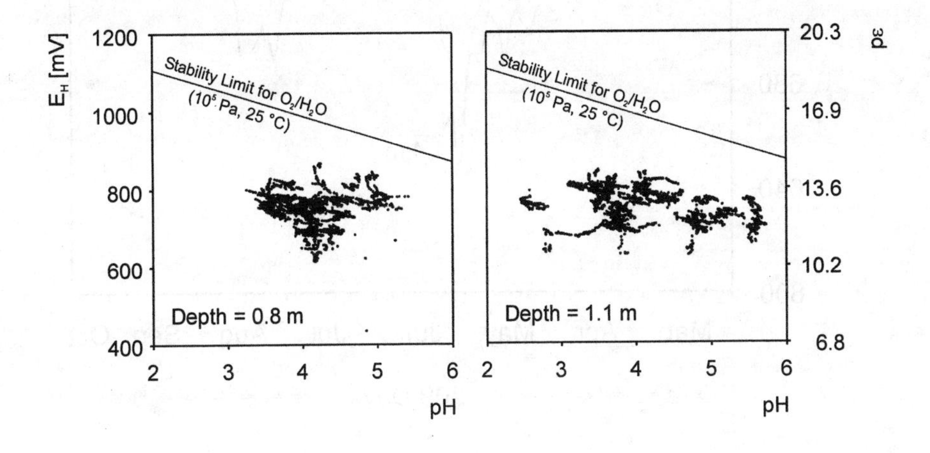

Fig. 8.2: E_H/pH- and pε/pH-diagram of measured redox potentials at the depth of 0.8 m and 1.1 m; redox potentials are the average of one day.

According to the classification and the pε/pH-diagram given in SPOSITO (1989) the redox state at a depth of 0.8 m and of 1.1 m throughout the measurement period must be classified as oxic, which means that oxygen reduction is the only possible redox reaction. This process is supposed to be responsible for the fluctuation of the redox potential up to approximately 250 mV (Figure 8.2), which was registered by the obviously sensible Pt-electrodes. However, from this fluctuation of redox potentials in oxic soil, no conclusions can be drawn with respect to a considerable solute and proton turnover in these depths, so that influences on soil acidity are not to be expected.

8.3.2
E_H at a Depth of 1.4 m

At a depth of 1.4 m only Pt-electrodes were installed at the beginning of the investigation period in March 1997. Therefore, only the temporal course of the redox potential is dealt with in the following.

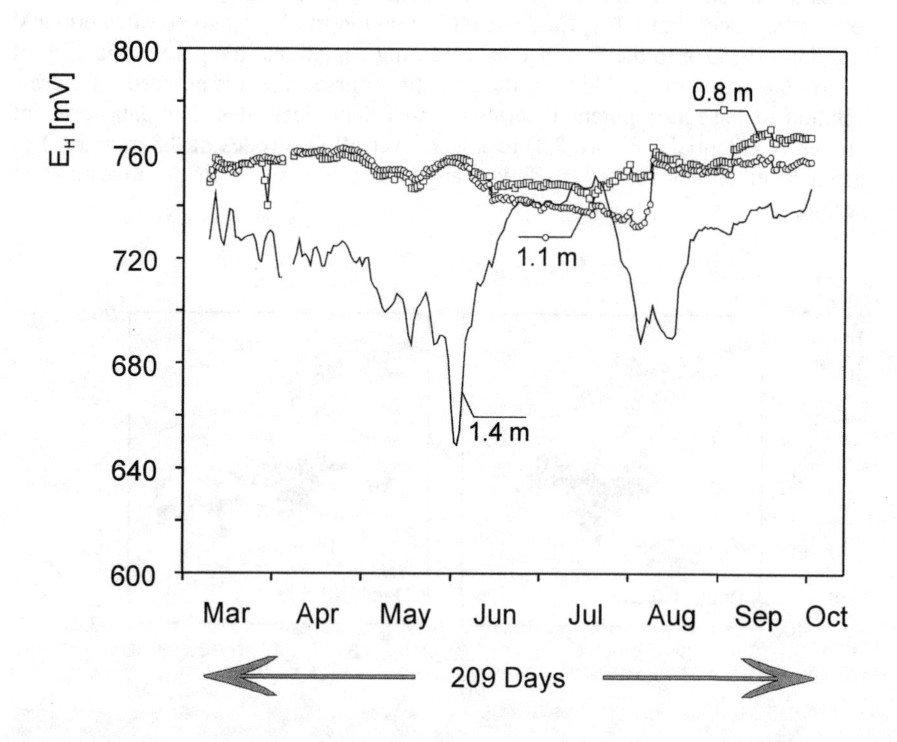

Fig. 8.3: Recorded course of redox potentials of a Gleyic Podzol in 1997; redox potentials are averaged over time and space (1 day and 19 replications).

In contrast to the redox potentials at a depth of 0.8 m and of 1.1 m, the redox potentials at 1.4 m depth (Figure 8.3) show considerably higher fluctuations ranging between 650 mV and 750 mV, but on average the redox state of the soil (SPOSITO, 1989) at a depth of 1.4 m must also be classified as oxic.

In May/June and August 1997 two clearly recognisable potential minima can be seen. Examining the recorded distribution of the redox potentials in time and space (Figure 8.4) it can be stated that the potential minima are mainly caused by the redox electrodes on measuring positions 2.5 m, 3.5 m, 5.5 m and 9 m. The E_H-values decreased temporally to 400 mV – 500 mV, on measuring position 5.5 m even to <50 mV, which points to suboxic or even anoxic conditions (SPOSITO, 1989). This demonstrates that the mere examination of potential series averaged over space, does not take potential fluctuations into consideration which are locally important. By using these data, an existing horizontal variability could not be detected. Because of the differing behaviour at the measuring positions, pH-electrodes and porous suction cups were installed additionally at positions 2.5 m, 3.5 m, 5.5 m and 9 m (see Figure 8.1). However, the evaluation of these is not available yet.

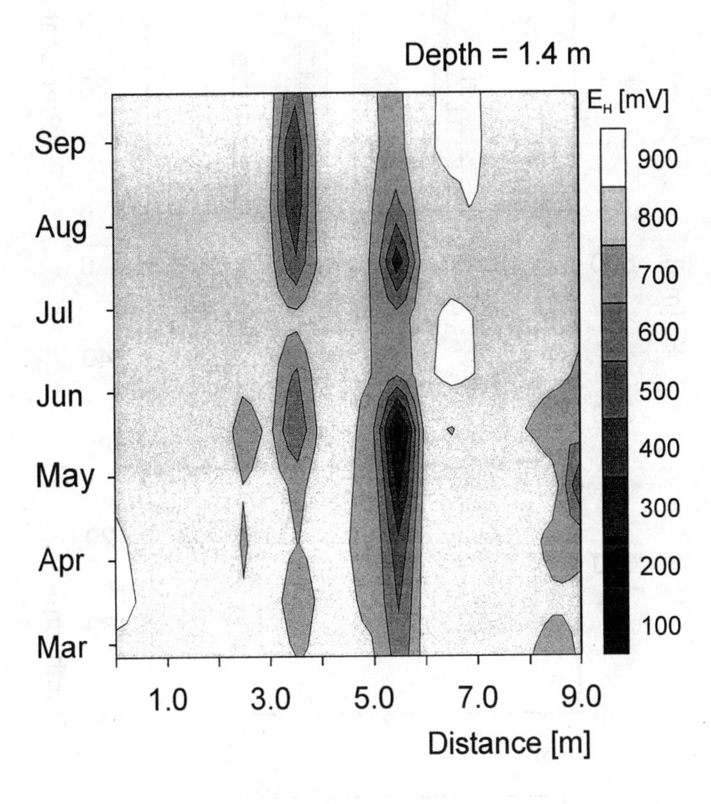

Fig. 8.4: Spatial and temporal structure of measured redox potentials at the depth of 1.4 m; redox potentials are the average of 1 day (contoured graph based on linear interpolation between single data).

Apart from the spatial variability of the redox potential, the temporal variation of the potential data series can be examined more closely at a measuring depth of 1.4 m. Consequently, at measuring positions 2.5 m, 3.5 m, 5.5 m and 9 m (Figure 8.4) it becomes obvious that after periods of strong potential decrease a quick return to E_H-values distinctly above 600 mV takes place. Long-term low redox

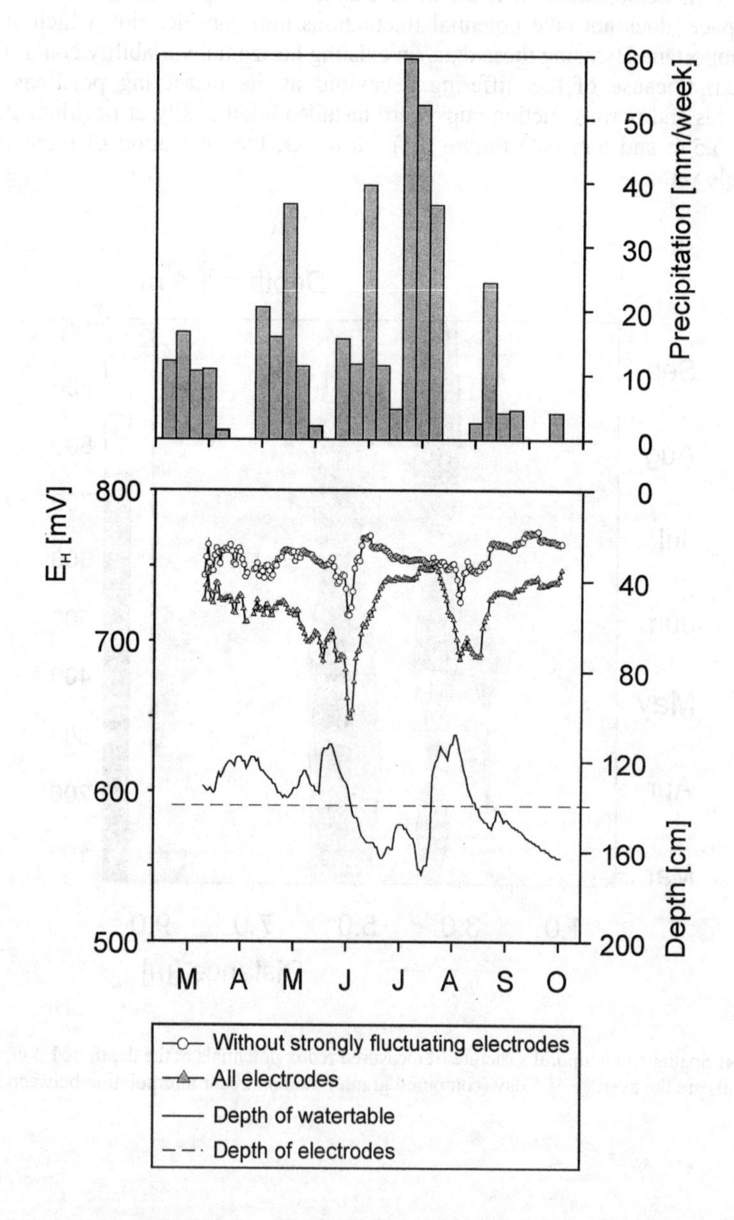

Fig. 8.5: Water table depth and measured redox potentials at the depth of 1.4 m (below), plotted data are average values of 1 day. The precipitation at the transect is shown above.

potential and therefore continuously reducing conditions cannot be observed at this depth either. Due to high temporal frequency of E_H-measuring, the phases of a strongly fluctuating redox potential can be recorded reliably. With a lower level of temporal frequency - or with electrodes only installed temporarily - it would not have been possible to detect the strong short-term potential fluctuations.

A comparison of the redox potential data series with the recorded groundwater table fluctuations reveals a contrary behaviour between the potential minima and the maxima of the groundwater amplitude (Figure 8.5). These results also remain unchanged when the strong fluctuating potential of the electrodes at depths 2.5 m, 3.5 m, 5.5 m and 9 m are not considered in the spatial mean of the data series (Figure 8.4). However, long-term water saturation (the Pt-electrodes were in the groundwater from March to the beginning of June) only leads to a decline of the potential by <100 mV with most of the electrodes (Figure 8.4). A distinct correlation between redox dynamics and groundwater dynamics was also found by MANSFELDT (1993), who measured redox potentials in the Go-horizon of a calcareous marshland soil by means of permanently installed Pt-electrodes. The recorded redox potentials in the marshland soil showed strong amplitudes of -180 mV up to 650 mV, which is an indication of intensive redox processes taking place in this young, fine-grained marshland soil. At our research site, the amplitudes of the redox potentials are much smaller, probably because the transport of oxygen in the sandy soil is not hampered by high water contents in the unsaturated parts of the soil profile.

Due to periods of rainfall excess (rainfall minus evapo-transpiration of the coniferous forest is positive) in the spring and summer months, the groundwater table rose temporarily clearly above the lowest redox measuring depth of 1.4 m, and afterwards decreased rather quickly (Figure 8.5). The long-term and short-term fluctuations in the redox potential and groundwater data series resulting from this, are to be considered as coincidental events in the time period involved. The temporal correlation between groundwater dynamics and mostly weak redox dynamics can be evaluated quantitatively by the spectral analysis of data series given in Figure 8.5.

The power spectra of data series of groundwater table and redox potential at a depth of 1.4 m (Figure 8.6) show significant peaks with maximum at f = 0.0117 [day^{-1}] , hereafter called f_{max}, which corresponds to a period of p = 85.5 days. Accordingly, in both data series a corresponding oscillation of the period p of about 86 days can be proven, which explains a very high proportion of the total variance of the data.

The agreement of the power spectra of the redox potential and the groundwater table is mainly traced back to 2 events (see Figure 8.5). It must be emphasised, that these oscillations are the product of randomly coincidental weather conditions. The proven periodicity of about 86 days is probably only valid for the measurement period. But throughout the measurement period the periodicity is rather pronounced as indicated by the rather narrow and highly significant peaks (Figure 8.6). Nevertheless, at the frequencies directly adjacent to f_{max} ($f_{max+1} = 0.0133\ day^{-1}$, $f_{max-1} = 0.0099\ day^{-1}$) the spectral power is in the same order of magnitude as at f_{max}, indicating that the period length of the oscillations may range from about 75 to about 100 days.

On examining the mean potential data series without the data measured at positions 2.5 m, 3.5 m, 5.5 m and 9 m (Figure 8.5), a significant peak which corresponds to p = 86 days can be stated too.

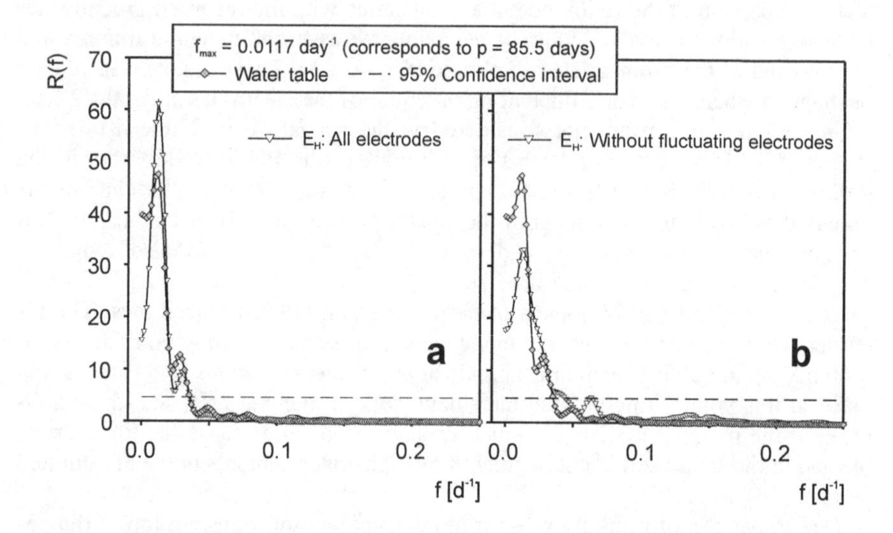

Fig. 8.6: Power Spectra of measured redox potentials (depth = 1.4 m) and water table depth. The redox potentials were averaged over all electrodes (a), averaged data measured without positions 2.5 m, 3.5 m, 5.5 m and 9 m (b). The 95% confidence interval refers to the significance of individual spectral power values (e.g. JENKINS & WATTS, 1968; SCHÖNWIESE, 1985).

In this adjusted data series an oscillation comparable with the groundwater periodicity is also inherent. Beside the peak at f = 0.0117 day^{-1} further small but significant peaks were discovered at f = 0.0283 day^{-1} (period p = 35 days) and at f = 0.0300 day^{-1} (period p = 33 days); i.e., apart from the obvious periodicity, both data series display a shorter, weaker oscillation of the time period from 33 to 35 days. This probably depends on short-term changes of weather conditions.

The amplitudes of the data series are not synchronous, the redox potential minimum follows the maximum peak of groundwater table with a delay of some days. In order to depict this delay without disturbing scatter of the data, the long-periodic oscillation of the groundwater table and the measured redox potentials (without the E_H-values at 2.5 m, 3.5 m, 5.5 m and 9 m) are presented in Figure 8.7 after FOURIER transform smoothing. The temporal delay of the E_H-minimum points out to reduction processes which are kinetically slow and microbially catalysed. On the basis of the mean oxic state of the soil and the small mean potential differences of <100 mV (Figures 8.5 and 8.7), this can be explained to be the result of mere respiration processes, which, in this measuring depth, - without any major solute and proton releases - lead to a reduction of partial pressure of oxygen. Only the temporally rather low redox potentials of individual measuring positions (especially at position 5.5 m) indicate conditions with a strongly reducing status (anoxic state). Here, reduction processes such as sulfate reduction cannot be excluded, which BÖTTCHER & STREBEL (1985) and STREBEL et al. (1990) clearly proved to take place in the aquifer at measured redox potentials of approximately 200 mV. So far, reasons for spatial redox-chemical heterogeneity of the investigated Gleyic Podzol and especially for the stronger amplitudes of the redox potentials at some positions in the 1.4 m depth are unknown. The studies on soil profiles in the neighbourhood of the measuring transect, however, indicate that at 1.4 m depth measurable organic carbon (C_{org}) contents with a variation coefficient of approximately 50% do exist (see Table 8.1). On the profiles, humus leaching down to a depth of 1.4 m (for example tongue-shaped, mottled) were visually perceptible. Furthermore, lenses of fine-grained material (mostly silt) could be found. Areas with higher C_{org}-contents and 'silt lenses' could have the effect of 'hot spots', where, spatially limited, more intensive redox processes take place, which in individual cases are confirmed by measurements with permanently installed electrodes. This conception about possible reasons for redox-chemical heterogeneity is supported by interpretations of COGGER et al. (1992) and MANSFELDT (1993), who traced back wide ranges of redox potentials, as found by in-situ measurements, to the existence of 'small-spatial' variability (microsites). Our measurements have been carried out in the undisturbed soil up to now. Therefore no information about the soil-chemical conditions at the individual measuring points is available yet.

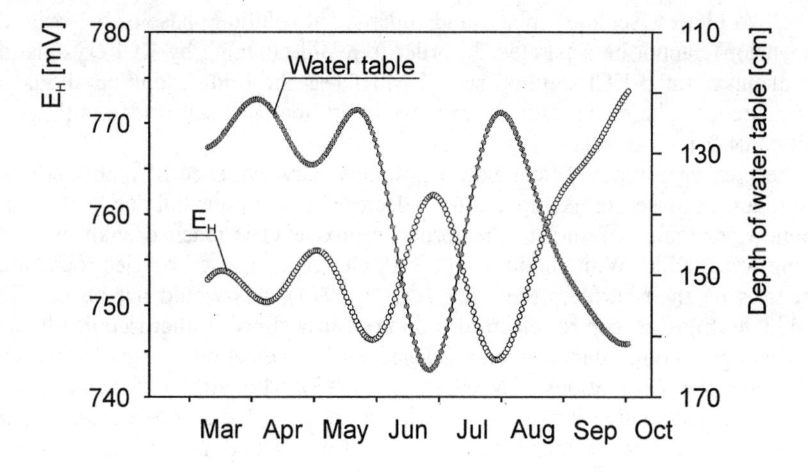

Fig. 8.7: Redox potential and water table after Fourier transformation smoothing.

In addition to these explanations related to the location, some fundamental reflections about potentiometrical measurements should be taken into account. CONKLING & BLANCHAR (1989) indicated that moisture gradients have an influence on potentiometrically measured pH. Increased variability was attributed to unreliable electrical contact between the reference system and the soil because of a decreased soil moisture content.

The reference electrodes used contain an electrolyte solution (3 M KCl), which is in contact - by the salt bridges - with the soil solution to serve as a passageway of electricity. If the kind and/or concentration of ions in the solutions are different, these ions would diffuse across the interface. Because of differences in diffusion rates of the charged ions, a potential difference between the two sides of the interface would result. This potential difference is called liquid-junction potential, or diffusion potential. Frequently, it is a source of error, particularly when measurements are made in colloidal systems, because traditionally it is assumed that the liquid-junction potential is small and constant in magnitude. In soil studies where the involvement of the effect of charged soil particles is frequently combined with a low (and changed) ionic strength, the liquid-junction potential represents a troublesome and complex phenomenon (BREZINSKI, 1983; YU & JI, 1993). At present, there is no way to overcome this problem entirely. Experimental conditions like the ionic strength, the pH and the temperature of the soil solution, affect the liquid-junction potential.

The extent of error which the mentioned factors have on the measured redox and pH-potentials is difficult to estimate, especially in case of measured redox potential fluctuations of only 50 to 100 mV. Laboratory experiments have to show the effect of moisture gradients on electrical contact. YU & JI (1993) made some suggestions for the use of reference electrodes in colloidal systems to reduce the liquid-junction potential. In field measurements, it is advisable to put the salt bridge on filter paper wetted by soil solution before. That is an alternative method, if the two phases (sediment phase and supernatant solution phase of the colloidal suspension) cannot be separated. In order to avoid clogging by KCl crystals, the use of unsaturated KCl solution such 3 M KCl as the bridge solution should be advantageous. These suggestions agree well with the used salt bridge outlined by VENEMAN & PICKERING (1983).

The measured redox potentials do not show any evidence of malfunctioning electrodes, because, for example, after a decrease of the potential due to the rising groundwater table, a return to the former redox level is reached again with the falling water table. With coatings on the Pt-electrodes caused by electrochemical processes on their surfaces (e.g. WHITFIELD, 1974), this would not be possible. Beside the transect, control electrodes do not show signs of microscopically visible coatings or other damages on the surface of the electrodes. This is in accordance with the observations of MANSFELDT (1993), who, after a 20-month-period of measurements taken in the Go-horizon and even in the sulfide-containing Gr-horizon of a calcareous marshland soil, could not notice any damages on electrodes.

8.4 Conclusions

For the characterisation of the redox state in acid sandy soils, Pt-electrodes proved to be a suitable tool for the long-term measurement of redox potentials. As the measurement results have shown so far, the temporal as well as the spatial variability of measured redox potentials in soils can be considerable. Therefore, E_H-measurements on a great number of positions are an inevitable prerequisite for examining the redox state of soils under field conditions. With the help of E_H-measurements, an identification of these parts of the soil volume is possible, for which specific additional investigations (soil solution, pH) are reasonable if redox processes are to be clarified. The groundwater fluctuations have a clear effect upon the redox potentials. However, the soil even below the groundwater table in the measured range is mainly oxic. Only locally and temporally anoxic conditions occur at a depth of 1.4 m. Therefore, it can be concluded that down to a depth of approximately 1.1 m, proton consuming redox processes, which can have an influence on soil acidity, are not of ecological relevance in this acid Gleyic-Podzol. The existence of low redox potentials at certain positions in the range of the fluctuating groundwater table (depth 1.4 m) make redox processes e.g. sulfate reduction locally likely. Because individual Pt-electrodes measure on selected points at a 'mm^3-volume', the quantitative importance and extension of such low redox potential zones ('hot spots') cannot be evaluated yet. Further measurements and specific evaluations must be carried out to determine whether these 'hot spots' are ecologically important for soil acidity and the composition of the groundwater recharge.

8.5 References

BÖTTCHER, J. & STREBEL, O. (1985): Redoxpotential und Eh/pH-Diagramme von Stoffumsetzungen in reduzierendem Grundwasser (Beispiel Fuhrberger Feld). Geologisches Jahrbuch, C 40: 3-34.

BÖTTCHER, J. & STREBEL, O. (1988a): Spatial variability of groundwater solute concentrations at the water table under arable land and coniferous forest. Part 1: Methods for quantifying spatial variability (geostatistics, time series analysis, Fourier transform smoothing). Z. Pflanzenernähr. Bodenk. 151: 185-190.

BÖTTCHER, J. & STREBEL, O. (1988b): Spatial variability of groundwater solute concentrations at the water table under arable land and coniferous forest. Part 3: Field data for a coniferous forest and statistical analysis. Z. Pflanzenernähr. Bodenk. 151: 197-203.

BÖTTCHER, J. & O. STREBEL (1990): Quantification of deterministic and stochastic variability components of solute concentrations at the groundwater table in sandy soils. pp. 129-140. In K. ROTH et al. (Eds.) Field-scale water and solute flux in soils. Birkhäuser, Basel, Switzerland.

BÖTTCHER, J.; LAUER, S.; STREBEL O. & PUHLMANN, M. (1997): Spatial variability of canopy throughfall and groundwater sulfate concentrations under a pine stand. J. Environ. Qual. 26: 503-510.

BREZINSKI, D.P. (1983): Influence of colloidal charge on response of pH and reference electrodes: the suspension effect. Talanta 30: 347-354.

COGGER, C.G.; KENNEDY, P.E. & CARLSON, D. (1992): Seasonally saturated soils in the puget lowland II. Measuring and interpreting redox potentials. Soil Sci. 154: 50-58.

CONKLING, B.L. & BLANCHAR, R.W. (1989): Glass microelectrode techniques for in situ pH measurements. Soil Sci. Soc. Am. J. 53: 85-62.

DEURER, M.; DUIJNISVELD, W.H.M.; BÖTTCHER, J. & EVERMANN, H. (1997): Die Struktur der räumlichen Variabilität der Wassercharakteristik als transportrelevanter Eigenschaft. Mitteilgn. Dtsch. Bodenkundlichen Gesellschaft 85: 75-78.

FRANKEN, G.; DUIJNISVELD, W.H.M. & BÖTTCHER, J. (1997): Eintrag von Acidität ins Grundwasser - Bestimmung der Saisonalität von Stoffflüssen unter Nadelwald (Fuhrberger Feld). Mitteilgn. Dtsch. Bodenkundlichen Gesellschaft 85: 717-720.

JENKINS, G.M. & WATTS, D.G. (1968): Spectral analysis and its applications. Holen-Day, San Franzisco.

KRUG, E.C. (1991): Review of acid-deposition-catchment interaction and comments on future research needs. Journal of Hydrology 128: 1-27.

MANSFELDT, T. (1993): Redoxpotentialmessungen mit dauerhaft installierten Platinelektroden unter reduzierenden Bedingungen. Z. Pflanzenernähr. Bodenk. 156: 287-292.

MARTH, C. & LOFTFIELD, N. (1991): Herstellung von Antimon-Mikroelektroden für pH-Messungen. Berichte des Forschungszentrums Waldökosysteme, Reihe B, vol. 24: 62-65.

MUELLER, S.C.; STOLZY, L.H. & FINCK, G.W. (1985): Constructing and screening platinum microelectrodes for measuring soil redox potential. Soil Science 139: 558-560.

PEIFFER, S.; KLEMM, O.; PECHER, K. & HOLLERUNG, R. (1992): Redox measurements in aqueous solutions - A theoretical approach to data interpretation, based on electrode kinetics. J. Contam. Hydrol. 10: 1-18.

PFISTERER, U. & GRIBBOHM, S. (1989): Kurzmitteilung : Zur Herstellung von Platinelektroden für Redoxmessungen. Z. Pflanzenernähr. Bodenk. 152: 455-456.

PUHLMANN, M.; BÖTTCHER, J. & DUIJNISVELD, W.H.M. (1997): Stoffkonzentrationsänderungen in Sand-Podsolen unter Kiefernforst bei unterschiedlichen Depositionssituationen: Ergebnisse aus einem Säulenversuch und Erklärungsansätze. Mitteilgn. Dtsch. Bodenkundl. Gesellsch. 85: 321-324.

SCHÖNWIESE, CH.-D. (1985): Praktische Statistik für Meteorologen und Geowissenschaftler. Gebrüder Bornträger, Berlin Stuttgart.

SIGG, L. & STUMM, W. (1996): Aquatische Chemie: eine Einführung in die Chemie wässriger Lösungen und natürlicher Gewässer. vdf-Hochschulverlag, Zürich.

SPOSITO, G. (1989): The chemistry of soils. Oxford University Press.

STREBEL, O.; BÖTTCHER, J. & FRITZ, P. (1990): Use of isotope fractionation of sulfate-sulfur and sulfate-oxygen to assess bacterial desulfurication in a sandy aquifer. J. Hydrology 121: 155-172.

STREBEL, O.; BÖTTCHER, J. & DUIJNISVELD, W.H.M. (1993): Ermittlung von Stoffeinträgen und deren Verbleib im Grundwasserleiter eines norddeutschen Wassergewinnungsgebietes. Umweltbundesamt, Texte 46/93, Berlin, Germany.

STUMM, W. & MORGAN, J.J. (1996): Aquatic chemistry: chemical equilibria and rates in natural water. 3rd ed., Wiley and Sons, New York.

TEICHERT, A.; BÖTTCHER, J. & DUIJNISVELD, W.H.M. (1997): Räumliche Variabilität bodenchemischer Eigenschaften am Beispiel eines Gley-Podsols aus Sand unter Kiefernwald. Mitteilgn. Dtsch. Bodenkundl. Gesellsch. 85: 369-372.

VAN BREEMEN, N. (1987): Effects of redox processes on soil acidity. Netherlands Journal of Agricultural Science 35: 271 - 279.

VENEMAN, P.L.M. & PICKERING, E.W. (1983): Salt bridge for field redox potential measurements. Commun. in Soil Sci. Plant Anal. 14(8): 669 - 677.

WHITFIELD, M. (1974): Thermodynamic limitations on the use of the platinum electrode in Eh measurements. Limnol. Oceanogr. 19: 857-865.

YU, T.R. & JI, G.L. (1993): Electrochemical methods in soil and water research. Pergamon Press, Oxford, New York, Seoul, Tokyo.

Chapter 9

Implementation of Redox Reactions in Groundwater Models

W. Schäfer

9.1 Introduction

Most groundwater constituents are subject to chemical or biochemical reactions during subsurface transport. A complete reproduction of the transport behaviour of such species in groundwater requires to take into account not only the physical transport processes advection and diffusion/dispersion, but also the balance of their transformations.

A straight way to consider redox reactions should be to use the redox potential as a so-called master variable which controls the distribution of redox-sensitive species. This approach requires that the modelled redox reactions proceed fast, i.e. that the time to reach thermodynamic equilibrium is short compared to the temporal resolution of the model, and that the reactions are fully reversible. However, only few redox reactions in the groundwater environment meet these two prerequisites. For instance, the oxidation of organic matter, which makes up an important reaction group in the subsurface, is usually an irreversible reaction. In such a case the redox potential can no longer be used as the exclusive governing parameter. Instead, non-equilibrium or kinetic approaches are necessary to simulate the redox reactions in a model.

The objective of this brief overview is to present and discuss examples for the different ways that exist for the incorporation of redox reactions in groundwater transport models.

9.2 The Redox Potential as Controlling Variable

9.2.1 Equilibrium Models

ENGESGAARD & KIPP (1992) presented a reactive transport model for the simulation of pyrite oxidation with nitrate as the oxidant (see Chapter 12 for a detailed review of this process). The overall reaction considered in the model can be written as:

$$5\,FeS_2 + 14\,NO_3^- + 4\,H^+ \Leftrightarrow 5\,Fe^{2+} + 10\,SO_4^{2-} + 7\,N_2 + 2\,H_2O \qquad (9.1)$$

The parameters and concentration values used in the model are adopted from a unconsolidated water table aquifer in Denmark. The objective of the model application was to reproduce the vertical movement of a redox front which divides the aquifer in an upper part where all pyrite has been oxidised, and a lower part which is still pyrite-bearing. As the authors aimed primarily at the vertical movement of groundwater and dissolved nitrate, they simplified the model by considering one-dimensional transport only. To account for the occurring redox reactions they coupled the geochemical model code PHREEQE (PARKHURST et al., 1980) to their transport model. The use of the coupled model allowed the reproduction of the slow downward movement of the redox front which is caused by the steady flux of nitrate dissolved in the recharging water.

The use of an equilibrium approach was possible as the pyrite oxidation proceeded fast compared to the vertical water and solute movement. This is also the reason for the formation of a sharp redox front. Although the reversibility criterion is not met (the formation of pyrite from a reaction with nitrogen gas and sulfate will not occur in aquifers), the equilibrium approach was successful, as the geochemical constraints were always such that the reverse reaction will not occur in the model. For those reasons, the redox potential could be used in this case as a master variable. Therefore, the knowledge of the redox potential is sufficient to calculate the distribution of all redox-sensitive species (problems connected with the experimental determination of the redox potential are discussed in Chapters 1 to 4).

However, problems may arise if geochemical conditions change in a way that the reverse reaction dominates in the model, whereas the reaction is actually irreversible in the subsurface. The following simple example illustrates this effect.

We will consider a hypothetical situation where acetate is oxidised with both oxygen and nitrate as the electron acceptors:

$$CH_3COOH + 2O_2 \Rightarrow 2H_2CO_3^* \qquad \Delta G_R^0 = -910\frac{kJ}{mol} \tag{9.2}$$

$$CH_3COOH + \frac{8}{5}NO_3^- + \frac{8}{5}H^+ \Rightarrow 2H_2CO_3^* + \frac{4}{5}N_2 + \frac{4}{5}H_2O \qquad \Delta G_R^0 = -889\frac{kJ}{mol} \tag{9.3}$$

Although these reactions are usually mediated by microbial metabolism in the environment, one could attempt to simulate them as purely chemical equilibrium reactions, if the kinetics are not of interest at the time. Figure 9.1 demonstrates the turnover that is calculated for an initial oxygen and acetate concentration of 0.31 mM and an initial nitrate concentration of 0.16 mM. It is assumed that the reaction occurs in the middle of the hypothetical model column (indicated by the arrow).

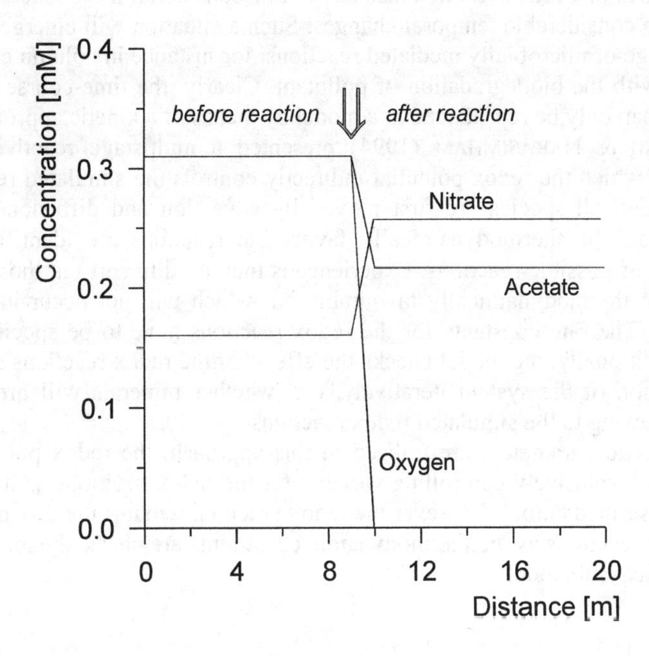

Fig. 9.1: Calculation of equilibrium concentrations for acetate, oxygen, and nitrate before and after hypothetical acetate oxidation.

As could be expected oxygen is nearly completely consumed and also part of the acetate is oxidised. Because excess acetate is provided in the model column another part of the acetate would be available for nitrate reduction. However, nitrate is not consumed but produced under the given boundary conditions. This means that Equation (9.3) proceeds in a reverse direction. These simulation results are not in agreement with the conditions given in subsurface environments, where acetate oxidation is irreversible, i.e. where the formation of nitrate and acetate from nitrogen, carbonic acid, and water can be excluded. The irreversibility of the reaction leads to erroneous results if an equilibrium approach is used. This error is easily detected for the simple example calculations shown in Figure 9.1, but it might be not so easy to identify artefacts arising from an inadequate application of an equilibrium model in more complex systems.

9.2.2 Combined Approach

If a redox reaction proceeds relatively slowly in the aquifer, sharp reaction fronts as shown in the example of the pyrite oxidation with nitrate will not develop, but the reaction zone will be distributed in a larger volume of the aquifer. The kinetics of a redox reaction also have to be considered if the reaction rates are subject to considerable temporal changes. Such a situation will emerge during the initial stage of microbially mediated reactions, for instance in column experiments dealing with the biodegradation of pollutant. Clearly, the time-course of a redox reaction can only be reproduced in a model by means of a kinetic approach.

McNab & Narasimham (1994) presented a multistage reactive transport model in which the redox potential indirectly controls the simulated reactions. In their model, all species are first moved by advection and diffusion/dispersion. Afterwards, the thermodynamically favourable reactions are identified from a given set of possible reactions. Experience is then used to sort out those reactions which are thermodynamically favourable but which will not occur in the actual situation. The rate constants for the redox reactions have to be specified by the user. Additionally, the model checks the effects of the redox reactions on the total composition of the system iteratively, e.g. whether minerals will precipitate or dissolve owing to the simulated redox reactions.

As kinetic parameters are utilised in this approach, the redox potential is no longer the exclusively controlling variable for the redox reactions as it was in the model described before. However the redox potential remains the driving force for the redox reactions as the thermodynamic constraints are checked explicitly in the geochemical submodel.

9.3 Models that do not Explicitly Consider the Redox Potential

9.3.1 Superposition Models

The superposition method presented here consists of three steps. First, the individual reactants of an anticipated redox reaction are moved independently. After the transport step the model checks the spatial overlap of the reactants. Finally, mass balance calculations based on stoichiometric relations between the different reactants are performed and the concentrations are updated according to these calculations. BORDEN & BEDIENT (1986) introduced this technique for a hypothetical in-situ bio-remediation of an aquifer contaminated with hydrocarbons (Figure 9.2).

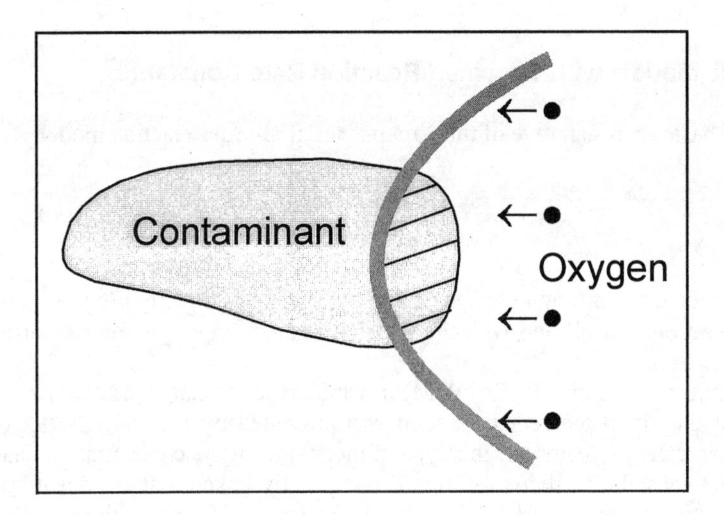

Fig. 9.2: Schematic representation of the superposition method for a hypothetical case of *in situ* bioremediation of groundwater.

For each transport time-,step the model verifies where dissolved oxygen and contaminant overlap (hatched area in Figure 9.2). In this zone, the oxidation of the contaminant is simulated by updating the contaminant concentrations according to the following calculation:

$$C_{new} = C_{old} - O \cdot f \quad (9.4)$$

where:

C_{new}, C_{old} = contaminant concentrations before and after the reaction step,
O = Oxygen concentrations before the reaction step,
f = stoichiometric factor of the contaminant oxidation.

The overlap and thus the specific decay rate generally increases for an increasing contaminant retardation compared to oxygen movement.

The superposition method is applicable if the kinetics of the redox reaction can be neglected. This may be true for microbial contaminant degradation if the degradation process holds up for a longer period of time until quasi steady-state conditions have evolved. In contrast to the equilibrium models presented before the superposition method accounts for the irreversibility characterising most of the microbially mediated redox reactions.

The superposition technique is not widely used for the simulation of redox processes. One reason might be that, in spite of the simple approach, the balancing method shown above tends to produce numerical instabilities. The redox potential is no longer a controlling variable. The redox state of the modelled system is only implicitly considered by the specification of the reactions that will be calculated.

9.3.2 Kinetic Models with Specified Reaction Rate Constants

The basic representative of this group is the first-order reaction model:

$$\frac{dC}{dt} = -\lambda C \tag{9.5}$$

Here the concentration change of a given species is directly proportional to the concentration C itself and with the reaction rate constant λ as the proportionality factor.

A multi-species model (CoTReM) in which reaction rate constants were specified for the simulated redox reaction was presented by LANDENBERGER (1998). One characteristic feature of this type of model is that the concentration change of a species caused by a redox reaction is not directly linked to the concentrations of the other reactants (see Chapter 11 for an example). Models with specified reaction rate constants are applicable in quasi steady-state conditions, i.e. if the species distribution in the model varies spatially, but only very slowly in time. This may be, for instance, true for marine sediments or pollution plumes from old spills with time-invariant boundary conditions. In contrast to the combined approach described above the redox potential no longer acts as a controlling variable. Instead, the user explicitly specifies the sequence of redox reactions under consideration. Inherent in this approach is the risk of specifying reactions which are thermodynamically not favourable.

9.3.3
Models with Variable Kinetics

Redox reaction kinetics in groundwater often exhibit appreciable spatial and temporal variability. For instance in case of degradation of the organic pollutants with different oxidants like oxygen, nitrate, and sulfate the effective degradation rates will depend on the availability of the respective oxidant. Usually this availability varies in different parts of the contaminated area. Typically, the interactions between the different degradation mechanisms create so-called redox-sequences in the downstream area of a contaminant source. Furthermore, the availability of the contaminants, or the different oxidants, will change in time for a certain position in the plume leading to time-dependent local degradation rates. It is rather awkward to reproduce such a variability in a model with prescribed reaction rate constants. A more convenient way to do this is to use a model where the rate constants automatically adapt to the geochemical constraints. An approved way to simulate the kinetics of microbially mediated redox reactions is to adopt so-called MONOD-kinetics. The functionality of this type of kinetic expression will be demonstrated for the aerobic degradation of an organic contaminant. First, the model computes microbial growth depending on the contaminant and oxygen concentrations:

$$\frac{dX}{dt} = \frac{C}{K_C + C} \cdot \frac{O}{K_O + O} \cdot v_{max} X \qquad (9.6)$$

where:

X = concentration of micro-organisms,
C = concentration of the contaminant,
O = concentration of oxygen,
v_{max} = maximum growth rate,
K_C, K_O = MONOD-constants for the contaminant and the oxygen, respectively.

Equation (9.6) describes an exponential microbial growth with a variable growth rate. Its value is maximal if both contaminant and oxygen concentrations are high. In this case both MONOD-terms are close to unity. If one of the concentration values or both of them are much smaller than the respective MONOD-constant, then the value of the MONOD-term will approach zero and microbial growth will cease. This mechanism ensures that aerobic micro-organisms do not grow outside the plume, where no contaminant is available, but also not in the centre of the plume, if oxygen is depleted there. Thus microbial growth is automatically restricted to the edges of the plume where contaminated waters mix with oxygen-rich pristine water.

Microbial growth and redox reactions are coupled via the yield coefficient Y and the stoichiometry factor F:

$$\frac{dC}{dt} = -Y \frac{dX}{dt} \qquad (9.7)$$

$$\frac{dO}{dt} = F\frac{dC}{dt} \tag{9.8}$$

Equations (9.6), (9.7) and (9.8) form a coupled system, i.e. a change in concentrations of the species feeds back on microbial growth. The effective degradation rates are now a function of the concentrations of all interacting species considered in the model. For instance a sequence of redox reactions downstream of a contaminant spill may be simulated by assigning different growth rates or MONOD-constants to the different microbial groups involved in the degradation process, or by using so-called inhibition terms which suppress the growth of e.g. sulfate reducers as long as dissolved oxygen is present.

As for the models with specified reaction rate constants the redox potential is not a controlling variable in the models with variable kinetics, but it is implicitly considered by the range and order of the modelled redox reactions. Again, the model user has to select the reactions carefully to avoid the simulation of reactions that are thermodynamically not favourable. Models with variable kinetics based on microbially mediated redox reactions are frequently used for the simulation of organic carbon degradation in sediments and aquifers (e.g. KINDRED & CELIA, 1989; LENSING et al., 1994; MACQUARRIE et al., 1990; SCHÄFER et al., 1998; WIDDOWSON et al., 1988).

9.4 Summary

It was not the objective of this chapter to provide a comprehensive review on redox modelling in aquifers but rather to highlight the different strategies employed to redox reactions in groundwater models. Emphasis was put on the significance of the redox potential for the simulated system.

It is only in equilibrium models that the redox potential is the exclusive controlling variable for the redox system. However, owing to the fact that many redox reactions in the subsurface proceed relatively slowly and that they are often irreversible equilibrium models are applicable to special situations and boundary conditions only, and mostly kinetic models have to be used instead. This is especially true for the simulation of redox reactions in connection with organic matter oxidation, which is a common field of application for reactive transport models. Consequently, the redox potential can no longer the be major controlling variable, and in many kinetic models it is not utilised at all. The redox reactions and their sequence have then to be specified extrinsically. Care must be taken to select reactions that are thermodynamically favourable. Generally, the redox potential is of minor importance for the simulation of redox reactions in the groundwater environment.

9.5 References

BORDEN, R.C., BEDIENT, P.B. (1986): Transport of Dissolved Hydrocarbons Influenced by Oxygen-Limited Biodegradation, 1. Theoretical Development. Water Resour. Res. 22: 1973-1982.

ENGESGAARD, P. & KIPP, K.L. (1992): A geochemical transport model for redox-controlled movement of mineral fronts in groundwater flow systems: A case of nitrate removal by oxidation of pyrite. Water Resour. Res. 28: 2829-2844.

KINDRED, J.S. & CELIA, M.A. (1989): Contaminant Transport and Biodegradation, 2. Conceptual Model and Test Simulations. Water Resour Res 25: 1149-1160.

LANDENBERGER, H. (1998): CoTReM, ein Multi-Komponenten Transport- und Reaktions-Modell. Berichte, Fachbereich Geowissenschaften, Universität Bremen, Nr. 100, Bremen.

LENSING. H.J.; VOGT, M. & HERRLING, B. (1994): Modelling of biologically mediated redox processes in the subsurface. J. Hydrol. 159: 125-143.

MACQUARRIE, K.T.B.; SUDICKY, E.A. & FRIND, E.O. (1990): Simulation of Biodegradable Organic Contaminants in Groundwater, 1. Numerical Formulation in Principal Directions. Water Resour. Res. 26: 207-222.

MCNAB JR., W.W. & NARASIMHAN, T.N. (1994): Modeling reactive transport of organic compounds in groundwater using a partial redox disequilibrium approach. Water Resour. Res. 30: 2619-2636.

PARKHURST, D.L.; THORSTENSON, D.C. & PLUMMER, L.N. (1980): PHREEQE - a computer program for geochemical calculations. U.S. Geological Survey.

SCHÄFER, D.; SCHÄFER, W. & KINZELBACH, W. (1998): Simulation of Reactive Processes Related to Biodegradation in Aquifers 1. Structure of the 3D Reactive Transport Model. J. Contaminant Hydrol. 31: 167-186..

WIDDOWSON, M.A.; MOLZ, F.J. & BENEFIELD, L.D. (1988): A numerical transport model for oxygen- and nitrate-based respiration linked to substrate and nutrient availability in porous media. Water Resour. Res. 24: 1553-1565.

Chapter 10

Variance of the Redox Potential Value in Two Anoxic Groundwater Systems

M. Kofod

10.1 Introduction

The use of redox potential measurements as an indicator of the redox milieu in geochemical environments is still subject of discussion. This chapter offers an empirical review of the redox potential measurements made in two groundwater systems over several years. The question is whether the measurements show statistically significant reproducibility so that the variance in values can be attributed to changes in the redox conditions. To maintain an empirical approach the correlation between E_H-values and the chemical composition as well as the spatial distribution of the redox potential measurements was tested. To this end, first the overall variance of the data was evaluated. In the second step variogram analysis was employed to determine how much of the total variance could be explained with the spatial variability of groundwater chemistry. This is followed by a correlation analysis to check if the measured redox potential is also statistically determined by the groundwater composition.

This empirical approach is performed on two sets of groundwater data (groundwater of the Oderbruch polder and of the river marsh in Hamburg). The data are obtained by routine groundwater monitoring. The redox potential measurement is part of field measurements. In both cases special measures to get more reliable redox data were not undertaken. Function tests and a preconditioning of the redox electrode prior to sampling were usually not performed. The redox potential was monitored during the purging of the wells. It was measured in a flow-through cell with commercially available redox electrodes (Ingold, Mettler). The redox potential value was read from the display of the voltmeter when slight changes of the value appeared ($\Delta mV < 10$ mV/10 min). Sampling was carried out by student workers (Hamburg) and technical staff (Oderbruch).

10.2 Groundwater in the River Marsh of the Elbe (Hamburg)

The aquifer of the marsh (Hamburg) is built up by 10 to 20 m sand. This sand is covered by the so called „Weichschichten“. These Holocene sediments are rich in clay and organic matter and have low hydraulic conductivity. In general, the aquifer is confined by these sediments. The hydraulic gradient in groundwater and the velocity of the groundwater flow are very low ($i < 0.001$, $v_f \approx 10$ m/a).

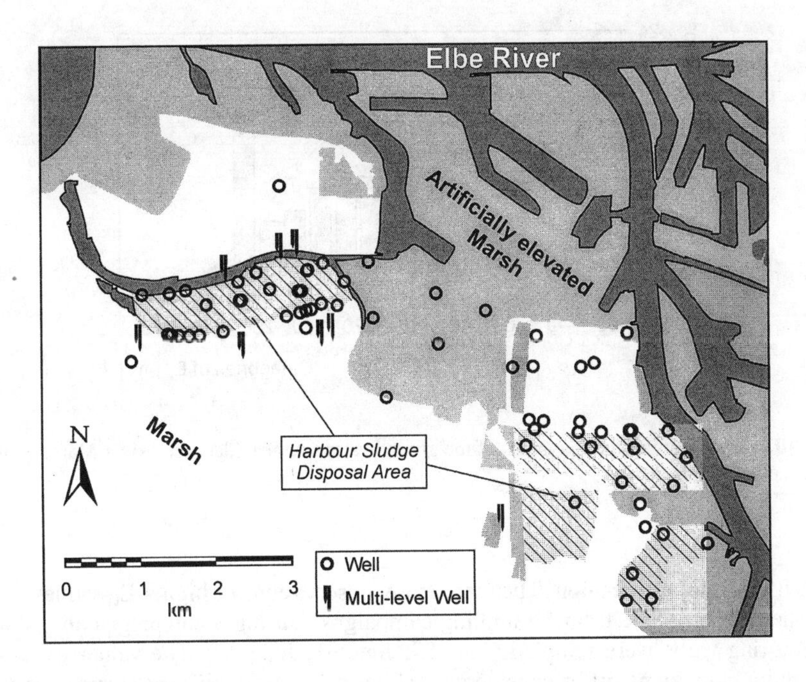

Fig 10.1: Distribution of the monitoring wells in the river marsh Hamburg (only wells of repeated E_H-value measurements are shown).

The composition of groundwater is determined by various processes (riverbed infiltration, input of pore water from the organic rich sediments and from the area of disposed harbour sludge). The groundwater composition reflects reducing conditions. The content of ammonium and iron is high. In some areas sulfate reduction can be observed. In a few of the monitoring wells influence of salt-rich groundwater appears that intrudes from a lower aquifer (GRÖNGRÖFT, 1992).

The monitored region covers an area of 60 km². The monitoring wells are mostly located near the harbour sludge disposal areas (Figure 10.1). The screen of the monitoring wells covers the complete aquifer. In addition to these wells, there are a few multi-level sampling devices installed for depth specific groundwater sampling.

10.2.1 Variation of the Measured E_H-Values

The measured redox potential* (E_H) of the groundwater ranges between +50 mV and -250 mV (Figure 10.2). The arithmetic mean is +116.7 mV with a standard deviation of ±51.2 mV. The histogram of the measured values does not exactly

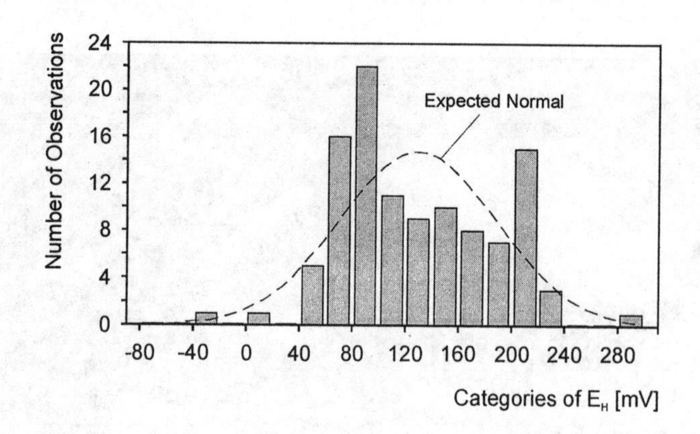

Fig. 10.2: Histogram of the E_H-values of the groundwater samples (data set "River Marsh Hamburg", all available data).

match a normal distribution. There are many observations of higher E_H-values.

The data originate from 5 sampling campaigns. During 3 campaigns almost all monitoring wells were sampled (June 89, June 91, June 92). The values of these sampling campaigns differ considerably (Figure 10.3). Significant changes of the composition of the monitored groundwater did not occur. The variance of the measured redox potential values over this period of time is an indication for irre-

* Data provided by Dr. A. Gröngröft, Institute of Soil Science, University of Hamburg.

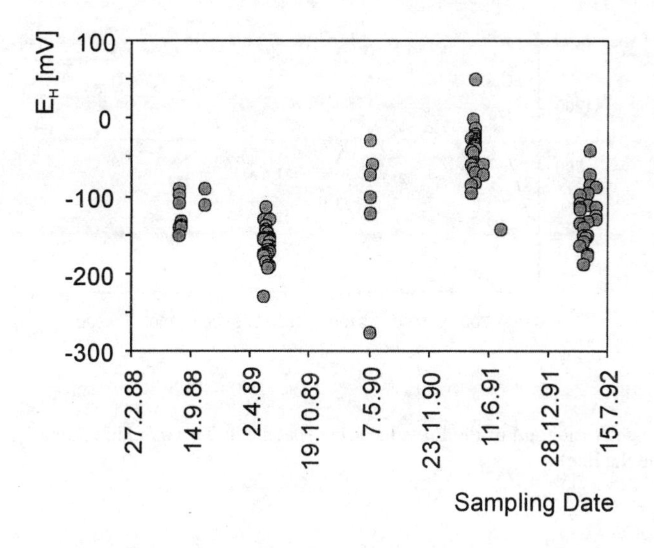

Fig. 10.3: Survey on the E_H-values of the individual sampling campaigns (data set "River Marsh Hamburg").

producible measurements. This could result from an alteration of the electrode characteristics due to wrong handling. It could also be an effect originating from different workers who have carried out the sampling procedure. Especially the results of the sampling in June 1991 seem to be outliers. Figure 10.4 shows the histogram after discarding the data of the 1991 sampling campaign. The histogram is very similar to a normal distribution with an arithmetic mean of +94.5 mV and a standard deviation of ±39.8 mV.

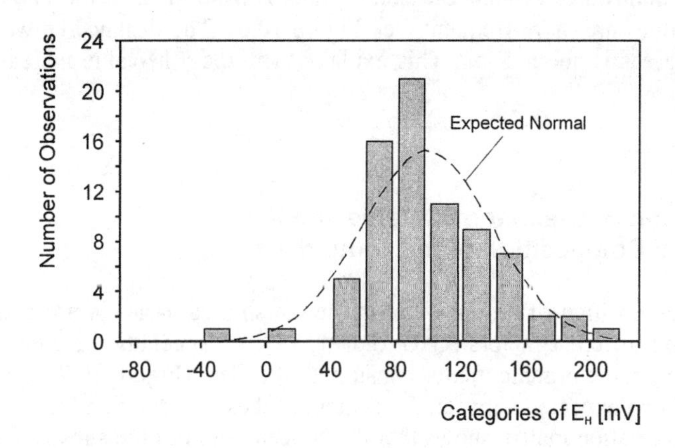

Fig. 10.4: Histogram of E_H-values without sampling campaign June 1991.

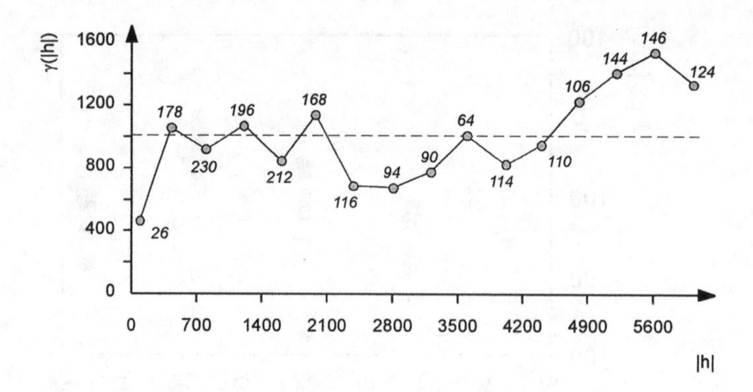

Fig. 10.5: Semivariogram of the E_H-values (lag spacing 400 m, omnidirectional, number of data pairs beside the line).

10.2.2 Spatial Distribution

Variogram analyses can be used to examine the spatial dependence of E_H-values. If a spatial dependence of the data exists, the value of γ(h) in a semi-variogram should increase with increasing distance between the measuring points (h) (JOURNEL & HUIJBREGTS, 1978). The semi-variograms of the groundwater data were produced with the program *VarioWin* (PANNATIER, 1996).

The semi-variogram of E_H-values does not show a steady increase of the γ(h)-value (Figure 10.5). From the first to the second lag (0 - 400 to 401 – 800 m interval) an increase of the γ(h)-value occurs. But the number of data pairs is very different (26 and 178). Therefore, this increase is not reliable. A significant increase of γ(h)-values appears after a distance of 5 km.

For groundwater around the south-eastern disposal area for harbour sludge higher values are more frequent (see Figure 10.6). The distance between the two disposal areas is about 5 km. This explains why the γ(h)-value increases beyond this range.

10.2.3 Correlation Between the E_H-Value and the Chemical Composition of the Groundwater

The composition of groundwater in the marsh area varies in a broad range depending on different factors (GRÖNGRÖFT, 1992). Especially the content of products of anaerobic organic matter consumption differs (Figure 10.7). So a relationship might be expected between the content of these products and the E_H-value.

The correlation matrix shows that the concentration of the substances are interrelated (Table 10.1). Most of these statistical relationships are very significant.

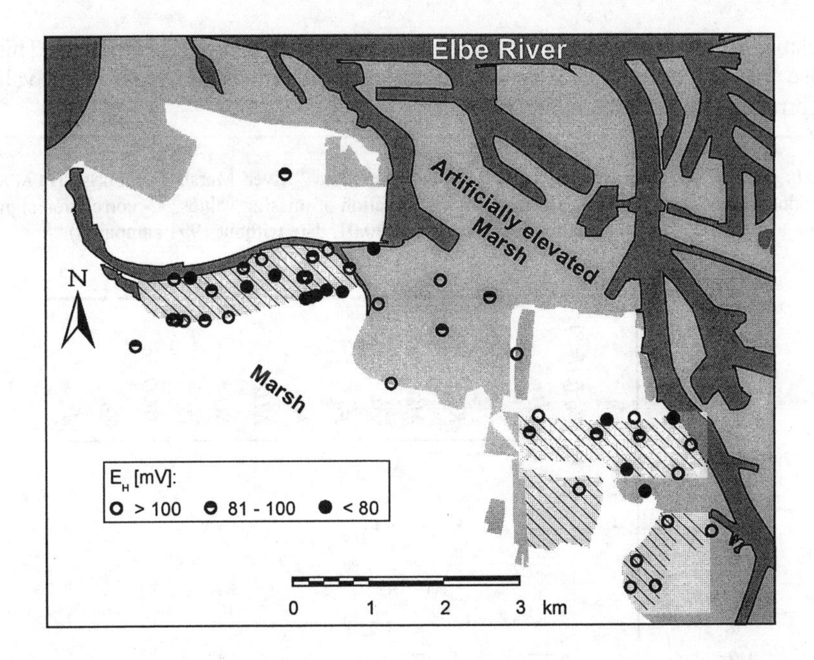

Fig 10.6: E_H-values of groundwater in the river marsh of Hamburg (data: sampling campaign 1988/89).

The E_H-value interrelates strongly with the pH (see Figure 10.8), the content of magnesium and hydrogen carbonate. These statistical relationships are congruent with the geochemical model on the processes which determine the groundwater composition in the monitored area (detailed description of the processes in GRÖNGRÖFT, 1992 and KOFOD, 1994).

Because of this significant correlation the E_H-value has been standardised to pH = 7 (Table 10.1). It would be expected that better correlation occur with the redox sensitive substances. The standardised E_H-value shows less significant cor-

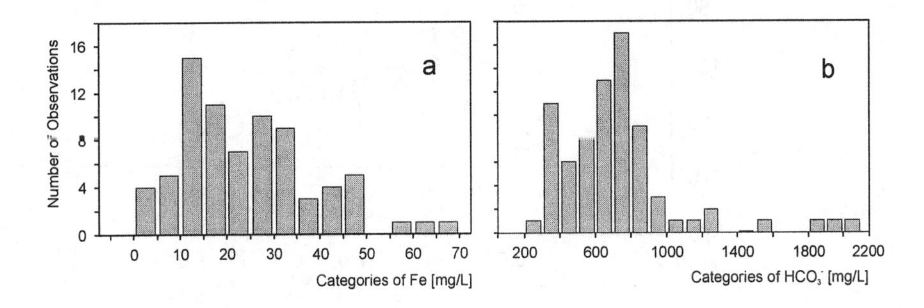

Fig. 10.7: Histograms of (a) iron and (b) hydrogen carbonate concentrations in the monitoring wells (data without 1991-campaign).

relation to the content of manganese, magnesium and hydrogen carbonate. This is also true when the concentration values are transformed into logarithmic values (correlation matrix not shown here).

Tab. 10.1: Correlation matrix of the groundwater data "River Marsh Hamburg" (PEARSON product-moment correlation, n ≈ 100, pairwise deletion of missing values, +/- correlation significant at $p < 0.05$, ++/-- correlation significant at $p < 0.01$, data without 1991 campaign).

Variable	E_H	E_H (pH=7)	pH	Na	K	Mg	Ca	Fe	Mn	NH_4	SO_4	Cl	HCO_3	Zn	As
E_H		--	--			--							-		
E_H (pH=7)	--														
pH	--				+		++				++				
Na												++			
K			+			++	++			++	++		++		
Mg	--	-	+		++		++	+	++	++	++		++		+
Ca	-		++		++	++			+	+	++		++		+
Fe						+			++	++			+		++
Mn		-				++	++	++		++	++		++		++
NH_4					++	++	++	++	++				++	+	
SO_4			++		++	++	++		++						
Cl				++											++
HCO_3	-	-			++	++	++	+	++	++			+		+
Zn										+					
As						+	+	++	++			++	+		

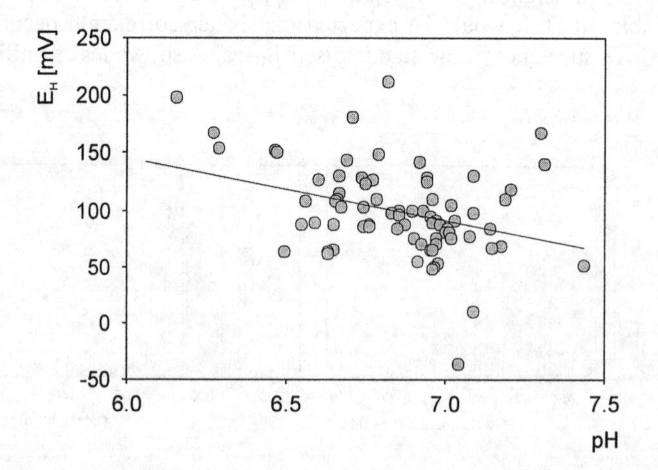

Fig. 10.8: XY-plot of pH and E_H-values (data set "River Marsh Hamburg", without sampling campaign 1991).

10.3
Groundwater in the Oderbruch (Brandenburg)

The Oderbruch (Land Brandenburg) is an artificially drained river plain next to the river Oder. The piezometric height of the groundwater lies below the water level of the river Oder. A dam prevents the river from flooding the Oderbruch. The aquifer of the Oderbruch has a thickness of 10 to 20 m and is built up by sand and gravel. In some parts the aquifer is confined by river sediments rich in clay and organic matter. In general the hydraulic gradient is small ($i < 0.001$). Near to the river Oder steeper gradients exist. Only in this small area a fast groundwater movement occurs. The chemical composition of the groundwater is determined by the infiltration of riverwater, by seepage and by lateral groundwater movement from higher areas in the south-west.

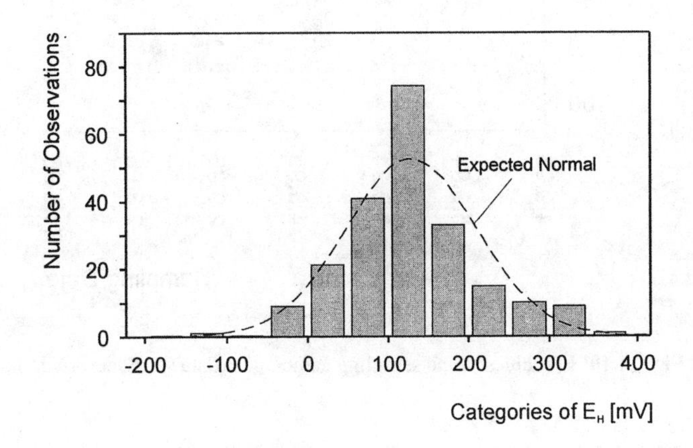

Fig. 10.9: Histogram of the E_H-values of the groundwater samples (data set "Oderbruch").

The area of the Oderbruch which is influenced by a lateral movement of groundwater from the south-west is not considered in the following evaluation.

The monitoring wells are distributed over the whole area of the Oderbruch (see Figure 10.13). In most cases the screen of the wells is located between 4 and 8 m below the surface and does not cover the aquifer completely. Therefore, the data only represent the composition of near surface groundwater.

The groundwater composition reflects reducing conditions as can be seen by the high iron content. Nitrate and oxygen do not appear in the groundwater of the Oderbruch (for further details see KOFOD et al., 1997).

10.3.1
Variation of the Measured E_H-Values

The measured redox potential* (E_H) varies between -200 and +400 mV. The arithmetic mean is +128.8 mV and the standard deviation is ±81.4 mV (n = 311). The histogram of the data demonstrates a normal distribution (Figure 10.9).

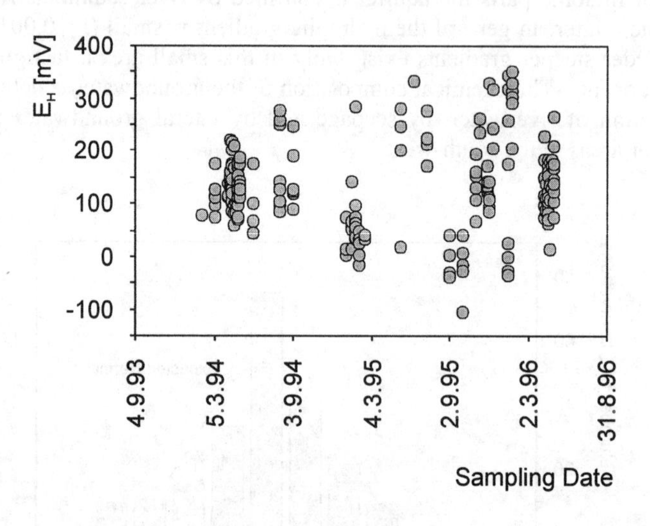

Fig. 10.10: E_H-values of the sampling campaigns (data set "Oderbruch").

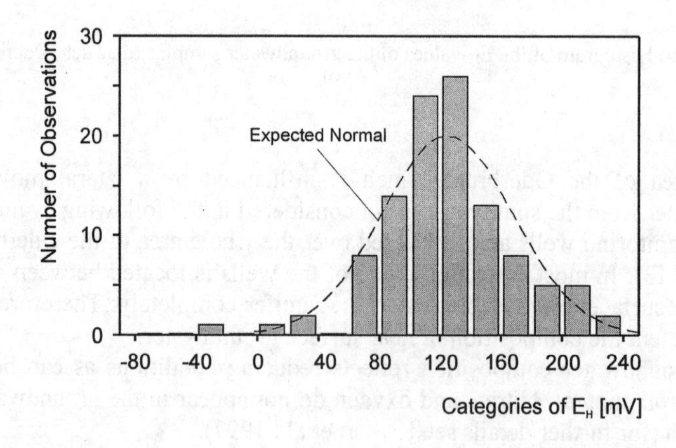

Fig. 10.11: Histogram of the E_H-values of the groundwater samples (data set "Oderbruch", without sampling campaign 1995).

* Data provided by Dr. C. Merz, Institute of Hydrology, ZALF e.V., Müncheberg.

The data originate from different sampling campaigns. As in the case of the data of the river marsh Hamburg, the measured E_H-values between these sampling campaigns differ (Figure 10.10). Significant changes of the groundwater composition are not likely. Again, this variation indicates irreproducible measurements. The individual evaluation of the data for each monitoring well showed that the data of the sampling during winter 95/96 are outliers. The histogram in Figure 10.11 is produced after discarding these data. The variance is significantly smaller (arithmetic mean = +94.5 mV, standard deviation ±39.8 mV).

10.3.2 Spatial Distribution

The variogram analysis was performed with the median of E_H-values (Figure 10.12). The semi-variogram of these data shows an increase of the $\gamma(h)$-value up to a range of 10 km. Beyond this range $\gamma(h)$-values vary unsteadily. The degree of the nugget-effect is 50 %. It can be assumed that a spatial dependence of the E_H-value exists.

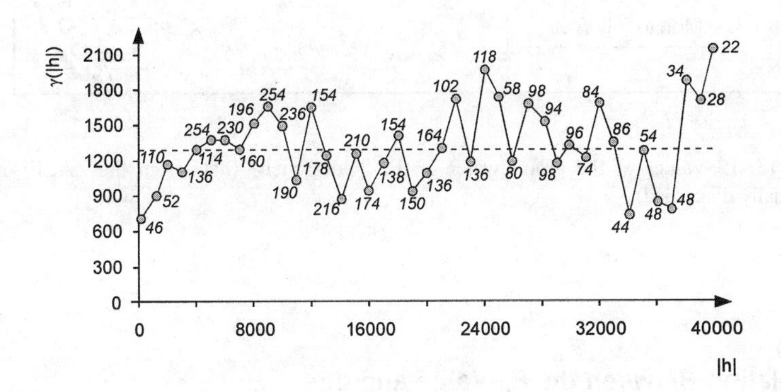

Fig 10.12: Semivariogram of the E_H-value of the groundwater date "Oderbruch" (lag spacing 1000 m, omnidirectional, number of date pairs beside the line).

In the map showing the groundwater E_H-values, one larger area with a steady change of E_H-values can be observed (Figure 10.13). In the northern part of the Oderbruch the E_H-values increase from the river Oder to the west. This can be explained by an increasing influence of seepage on the chemical composition of the groundwater. It has to be kept in mind that the screens of the monitoring wells do not cover the complete aquifer. The monitoring wells are very sensible to changes in groundwater composition induced by seepage water. More irregular distributions of the E_H-values occur in the southern part of the Oderbruch.

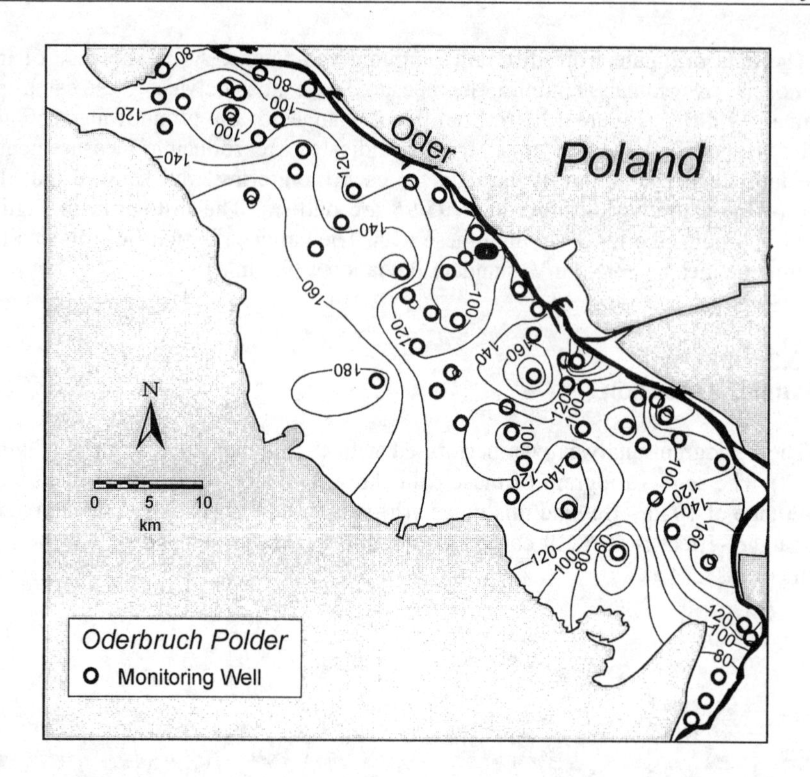

Fig 10.13: E_H-values of the groundwater in the "Oderbruch" (data from the ZALF, outliers individually discarded).

10.3.3 Correlation Between the E_H-Value and the Chemical Composition of the Groundwater

The composition of groundwater is determined by different processes (river bed infiltration, substance input by seepage, redox processes during the lateral flow, see KOFOD et al., 1997). Because of the different factors the groundwater composition varies in a wide range. In contrast to the groundwater of the river marsh in Hamburg, the input of substances by means of seepage is an additional factor. This is indicated by the broad variation of sulfate concentrations (Figure 10.14).

The contents of the different substances in the groundwater are strongly interrelated (Table 10.2). The E_H-value correlates most significantly with the pH. A correlation with the content of magnesium also exists. The linear regression between the pH and the E_H-value has a slope of 89.4 mV/pH (Figure 10.15). It is therefore assumed that the variance of the E_H is not only determined by the variance of pH alone. The standardised E_H-value correlates with the pH, the content of ammonium and hydrogen carbonate, and more significantly with the content of manganese.

Tab. 10.2: Correlation matrix of the groundwater date "Oderbruch" (PEARSON product-moment correlation, $n \approx 130$, +/- correlation significant at $p < 0.05$, ++/-- correlation significant at $p < 0.01$, data without 1991 campaign).

Variable	E_H	E_H (pH=7)	pH	Na	K	Mg	Ca	Fe	Mn	NH_4	SO_4	Cl	HCO_3	Zn	As
E_H		++	--			+									
E_H (pH=7)	++		-					--		-			-		
pH	--	-				--	--	--			--			-	
Na						++						++			
K						++							++		
Mg	+		--	++	++		++	++	++		++	++	++	+	
Ca			--			++		++	++		++	+	++		
Fe			--			++	++		++	++	++		++	++	
Mn		--				++	++	++		++	++		++		
NH_4		-						++	++				++		
SO_4			--			++	++	++	++			++	++		
Cl				++		++	+				++				
HCO_3		-			++	++	++	++	++	++	++			++	
Zn			-			+		++					++		++
As														++	

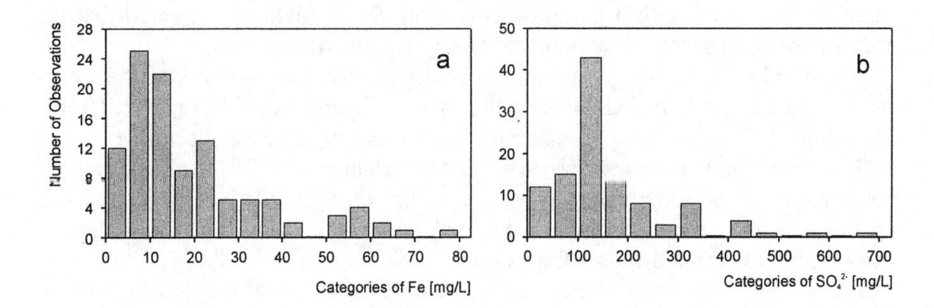

Fig. 10.14: Histograms of (a) iron and (b) sulfate concentrations in the monitoring wells of the "Oderbruch" (data without sampling campaign 1995).

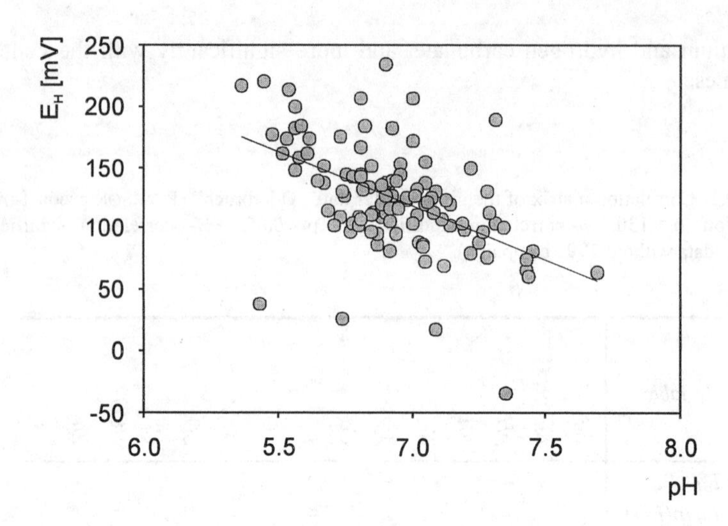

Fig. 10.15: XY-plot of the pH and E_H-values in the Oderbruch polder (without sampling campaign 1995).

10.4 Interpretation and Conclusion

In general, the reproducibility of "good" redox measurements is assumed to be in the range of ±50 mV (KÖLLING, 1986). Compared to this value, the variances of both data sets are small. The standard deviations of the complete data are ±51.2 mV (river marsh Hamburg) and ±81.4 mV (Oderbruch polder). One part of this variance seems to be caused by systematic differences between the results of the sampling campaigns. Discarding the results of the most differing sampling campaign decreases the standard deviation of the remaining data (±39.8 mV: River Marsh Hamburg, ±43.3 mV: Oderbruch). The histogram of both data sets matches quite well with a normal distribution. So it might be expected that the data variance is caused by random and not by systematic effects.

Nevertheless the variogram analysis indicates that the variance of the Oderbruch data can partly be explained with a spatial dependence. The nugget effect is quite high (50%) but the $\gamma(h)$-value of the variogram increases up to a range of 10 km. Mapping the data shows that a steady change of the E_H-value occurs in some areas of the Oderbruch. There are some uncertainties but in general this spatial dependence corresponds to the explanatory model on the type of factors determining the groundwater composition.

It is not sure whether a spatial dependence of E_H-values also exists in the case of the river marsh in Hamburg. The $\gamma(h)$-value of the semi-variogram increases from the first to the second lag, but the data pairs for calculating the $\gamma(h)$-value differ strongly. Mapping shows that the E_H-values in some areas are more frequent than in other. To be sure that this really is a spatial dependence, an intensive evaluation of the groundwater composition data and the hydraulic setting is required.

The correlation analysis indicates that the E_H-value of groundwater in the river marsh of Hamburg depends statistically on the pH and on the content of calcium, magnesium and hydrogen carbonate. The linear regression of the XY-plot shows a pH-dependence with a slope of ≈55 mV/pH. This is what theoretically could be expected if only pH determined the variation of the E_H-value. After standardising the E_H-value to a pH of 7 only less significant correlations remain.

The E_H-values of the Oderbruch data set are also significantly correlated with the pH-values but the slope is much steeper than 56 mV/pH. The linear regression is described by the function

$$E_H \text{ [mV]} = -89.4 \text{ pH} + 744 .$$

The differences in the E_H-value seem not to be determined exclusively by the pH alone. The standardised E_H-value for pH = 7 correlates with ammonium and hydrogen carbonate and very significantly with the content of manganese.

Table 10.3 gives some empirical information concerning E_H-values which are typically observed under the different reducing conditions. The reducing conditions are here characterised by the dominant e^--acceptor.

Tab. 10.3: Empirical findings on upper E_H-boundaries of different redox reactions (E_H-values in [mV]).

	Soil			*Groundwater*	
Electron acceptor	BOHN, 1971	BRÜMMER, 1974	SCHEFFER, 1979	BÖTTCHER & STREBEL, 1985	DVWK, 1988
NO_3^-	< 225		450 to 550	300 to 500	
Mn(IV)	< 200	< 450	300 to 450		
Fe(III)	120 to 280	< 220	< 150		200 to 300
SO_4^{2-}	-150 to 100	< 10	< -50	150 to 280	

The groundwater composition of both investigated areas are influenced by the anaerobic consumption of organic matter. Most obvious is the high content of iron induced originating from iron reduction. Sulfate reduction is of minor importance and is only observed in a few monitoring wells. The arithmetic mean of the E_H-value is +95 mV (river marsh Hamburg) and +125 mV (Oderbruch polder). Relating to Table 10.3 the height of these E_H-values seem to match well.

It can be summarised that E_H-measurement is not obsolete, because of a too low reproducibility. Even in the case of these two data sets with low total variance

(<50 mV, after discarding statistical outliers) systematic dependencies between the E_H-value and the chemical composition and the spatial variability of the groundwater can be observed. This result should encourage continue with the redox measurements and to improve the results by optimising the handling of the redox electrodes.

10.5 Acknowledgements

The author thanks Dr. Alexander Gröngröft (Institute of Soil Science, University of Hamburg), Dr. Christoph Merz and Dipl. Geol. Ingo Siekmann (Institute of Hydrology, ZALF Müncheberg) who provided the data of the groundwater composition in both areas. I like to thank Prof. Margot Isenbeck-Schröter and Dipl. Geol. Verena Haury for their assistance in the preparation of this manuscript.

10.6 References

BÖTTCHER, J. & STREBEL, O. (1985): Redoxpotential und Eh/pH-Diagramme von Stoffumsetzungen in reduzierendem Grundwasser (Beispiel Fuhrberger Feld). Geologisches Jahrbuch Reihe C Heft 40, E. Scheizerbart´sche Verlagsbuchhandlung , Stuttgart.

BOHN, H. L. (1970): Redoxpotentials. Soil Science 112: 39-45.

BRÜMMER, G. (1974): Redoxpotentiale und Redoxprozesse von Magan-, Eisen- und Schwefelverbindungen in hydromorphen Böden und Sedimenten. Geoderma 12: 207-222.

DVWK-Schriften (1988): Bedeutung biologischer Vorgänge für die Beschaffenheit des Grundwassers. Schriftenreihe des Deutschen Verbandes für Wasserwirtschaft und Kulturbau e.V. Heft 80, Paul Parey, Hamburg Berlin.

GRÖNGRÖFT, A. (1992): Untersuchung des Sickerwasser- und Stoffaustrages aus Hafenschlick-Spülfeldern in den oberen Grundwasserleiter der Hamburger Elbmarsch. Hamb. Bodenkdl. Arbeiten 17, Verein zur Förderung der Bodenkunde in Hamburg, Hamburg.

JOURNEL, A.. G. & HUIJBREGTS, C. J. (1978): Mining Geostatistics. Academic Press. London.

KOFOD, M. (1994): Die Bedeutung frühdiagenetischer Prozesse für die Porenwasserzusammensetzung in anaeroben Baggerschlämmen. Hamb. Bodenkdl. Arbeiten 28, Verein zur Förderung der Bodenkunde in Hamburg, Hamburg.

KOFOD, M.; SCHÜRING, J.; MERZ, C.; WINKLER, A.; LIEDHOLZ, T.; SIECKMANN, I. & ISENBECK-SCHRÖTER, M. (1997): Der geochemische Einfluß von Sickerwasser aus landwirtschaftlich genutzten Flächen auf das Grundwasser im Oderbruch. Z. Dt. Geol. Ges. 148: 389-403.

KÖLLING, M. (1986): Vergleich verschiedener Methoden zur Bestimmung des Redoxpotentials natürlicher Wässer. Meyniana 38: 1-19.

PANNATIER, Y. (1996): VarioWin: Software for Spatial Data Analysis in 2D. Springer Verlag, Berlin Heidelberg New York Tokio.

SCHEFFER, F. (1979) Lehrbuch der Bodenkunde / Scheffer-Schachtschabel. Enke Verlag, Stuttgart.

Chapter 11

Redox Fronts in Aquifer Systems and Parameters Controlling their Dimensions

J. Schüring, M. Schlieker & J. Hencke

11.1 Introduction

Most hydrochemical systems are to a large extent characterised by redox-processes which may lead to secondary reactions as a result of changing geochemical equilibria. They depend on the abundance of oxidising and reducing agents. Driven by the oxidation of organic substance which may either be dissolved in the liquid phase or bound to solids, a typical sequence of events in the consumption of oxidising agents is determined by the decreasing energy budget along the reaction path from the oxic zone into the zone of methanogenesis.

Whereas the reduction of oxidised manganese and iron minerals depends on their physico-chemical availability, the consumption of the dissolved species oxygen, nitrate and sulfate is only controlled by their fluxes and the consumption rate of organic matter. Redox processes in the aquatic environment are generally mediated by microbial activity reflecting the thermodynamical energy yield. Hence a sequence of characteristic micro-organisms can be found from heterotrophic oxygen consumers, denitrifiers, Mn- and Fe-reducers, sulfate reducers to methane fermenting organisms. The capability of these bacteria to oxidise organic carbon depends on the environmental conditions (e.g. salinity or temperature) as well as the bio-availability of organic carbon. As a consequence, organic carbon is in most cases partially rather than completely consumed.

The term reaction front describes the moving zone in which reactions take place. Its velocity is controlled by the consumption rates. For redox controlled reactions this zone is characterised by the consumption of electron donors and acceptors. During the establishment of a reaction front we have to distinguish two mechanisms which may be described by non-steady-state and steady-state conditions, respectively. This may be illustrated with the help of a beaker filled with organic-rich sediments: The whole system is water saturated, hence the organic matter may be oxidised by dissolved oxygen. It is possible to measure the oxygen distribution within the sediment by means of micro-electrodes. Immediately after filling the beaker we have non-steady-state conditions, which means that there is no equilibrium between the velocity of oxygen transport into the sediment and the oxidation rate of organic carbon. This leads to an increasing oxygen penetration depth over time until the steady-state between oxygen flux and oxidation rate has been established. From this time onward the steady-state condition is only controlled by the amount of organic carbon in the sediment. Once the organic matter is depleted, the reaction front moves downward into the sediment. In case of diffusive oxygen transport, the concentration gradients decrease with increasing depth. Therefore the velocity with which the front moves slows down with depth. In the case of advective transport and constant reaction rates, the front would move at a constant velocity along the flow-path.

In the case that the amount of bio-available organic carbon is not limited with respect to possible redox reactions, the electron acceptors oxygen, nitrate, oxidised Mn- and Fe-phases and sulfate are subsequently reduced. In this case, a reaction front is established by the depletion of oxidised manganese and iron minerals in the solid phase. The reductively released iron may be immobilised again to a variable extent as sulfate reduction becomes the process favoured by micro-organisms resulting in a release of sulfide. The reaction front moves in time together with the ongoing depletion and enrichment, respectively, and is marked by variations of the redox potential.

In many hydrochemical environments, the sequence described by BERNER (1981) is not complete because the amount of electron donors is limited in the solid phase or because the system lacks dissolved organic carbon. LOVLEY & GOODWIN (1988) postulate on the basis of hydrogen concentrations in an aquifer system that varying partial pressures of hydrogen are inducing the establishment of characteristic microbial populations. Sulfate-reducing organisms should therefore be active at higher energy states compared to iron reducing bacteria (LOVLEY & PHILLIPS, 1987). In systems which are rich in ferric iron, the activity of sulfate-reducing organisms is therefore inhibited by iron reducing bacteria (CORD-RUWISCH et al., 1988). This is typical for most groundwaters where a sulfidic environment is very rarely observed. In this case, the redox sequence is either incomplete or extends over long distances (LOVLEY & PHILLIPS, 1987). In marine or limnic sediments the sequence is mainly controlled by the sedimentary input of organic substance the reactivity of which may be variable. Other authors (e.g. POSTMA & JAKOBSEN, 1996) discuss the fact, the redox zones are to a lesser extent a result of the energy yields than they display various states of the establishment of an equilibrium. This could be one explanation for the spatially simultaneous reduction of ferric iron and sulfate as a result of different reaction kinetics (POSTMA & JAKOBSEN, 1996).

The redox environment influences a wide range of ions. The investigation of redox sensitive metals has a long tradition particularly in the marine sciences. Sudden changes of the depositional environment which may be induced by Pleistocene climatic changes have induced strong variations in the redox conditions due to changing sedimentation rates and varying paleo-productivities. Many redox sensitive trace elements such as vanadium, chromium, arsenic, molybdenum and uranium serve as proxy parameters for palaeo redox-conditions (CALVERT & PEDERSEN, 1993; FRANCOIS, 1988).

11.2 Regional Scales of Redox Fronts in Aquifer Systems

The geochemical conditions of groundwaters are established to a major part during groundwater recharge on the way through the unsaturated zone (BRADLEY et al., 1992). WENDLAND et al. (1994) postulated that approximately 50 percent of the anthropogenic nitrate in Germany is reduced in the unsaturated zone. Decaying organic substances are responsible for the continuous input of organic carbon. In the unsaturated zone oxygen is the most important oxidising agent. In contrast to groundwater conditions, the abundance of oxygen is much higher in the unsaturated zone. This is mainly due to higher diffusion coefficients in the gas phase and the higher concentration gradient with respect to the partial pressure of the atmosphere. In addition, barometric effects as well as temperature changes may lead to convective gas exchanges. It is obvious that the extent of groundwater recharge depends on the hydraulic conductivity of the unsaturated zone as well as the vegetation density. With respect to the geochemical imprint this means that the major amount of groundwater is recharged in areas with sparse vegetation and in areas with highly conductive soils. Since the extent of redox processes depends on the abundance of organic matter as well as the residence time, the redox conditions of most groundwaters do not permit sulfate reduction or even methanogenesis. Strongly reducing conditions are therefore characteristic for low conductive fine-grained soils through which groundwater recharge is often negligible or in swampy areas.

Redox processes taking place in the saturated zone are controlled by the abundance of electron donors and acceptors. The residence times are much higher in aquifers compared to the time seepage water needs on its way to the groundwater. In addition, the composition of the solid phase is controlled by the sedimentary environment during deposition of the material. Hence, in comparison with marine or limnic systems, where the solid phase is permanently added by sedimentation, aquifers represent a fixed pool system with respect to geogenic solid phases. Reaction fronts in general may establish due to reactions of dissolved species transported by the groundwater flow leading to characteristic depletions and/or enrichments of solid phases. The comparatively low reduction capacity of most aquifers is illustrated by the fact that nitrate from intense agriculture is found in almost all groundwaters where its occurrence in deep groundwaters is an increasing problem for waterworks. This is especially important in hard rock aquifers were the abundance of organic carbon is solely restricted to the input via groundwater recharge.

The limited reducing capacity of pore-water aquifers is well illustrated by the Tertiary sediments of Middle Europe. Although the groundwater residence times are estimated to be more than 1000 years, most groundwaters are characterised by high nitrate concentrations. Since the amount of reactive organic carbon is low, several authors discuss the possibility of nitrate reduction by oxidation of synsedimentary iron sulfides which may explain the increased sulfate concentrations in that region (e.g. KÖLLE et al., 1983; POSTMA et al., 1991; VAN BEEK & VAN DER KOOIJ, 1982; VAN BEEK, see Chapter 12; VAN BEEK et al., 1988).

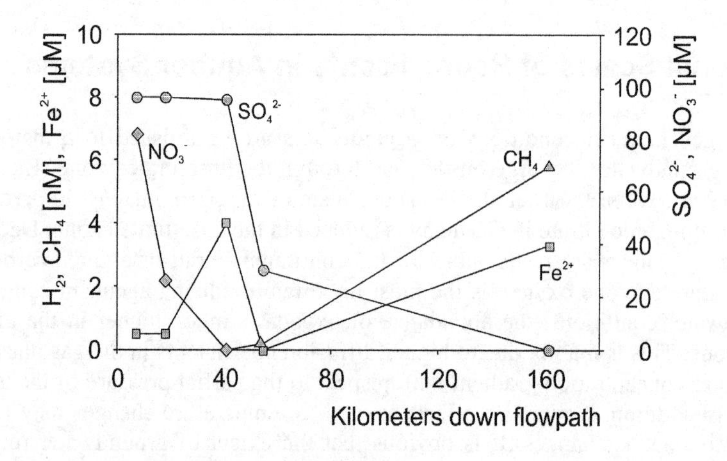

Fig. 11.1: Redox succession as observed by LOVLEY & GOODWIN (1988, modified) in the *Middendorf* aquifer, South Carolina.

Laterally proceeding sequences of redox processes along the flow-path are rare and generally restricted to certain hydraulic conditions such as bank filtration. In case of low groundwater velocities, a vertical sequence of redox sequences may develop as a result of groundwater recharge. LOVLEY & GOODWIN (1988) observed a complete redox sequence over a distance of 150 kilometres (Figure 11.1).

11.3 Investigating the Scales of Redox Fronts

11.3.1 Column Tests

Laboratory experiments offer the easiest way to study the development of redox fronts. The example shown here bases on results from column experiments using natural aquifer material. The silty sands were built in columns with a length of 50 centimetres and a diameter of 4.2 centimetres. The sediments are characterised

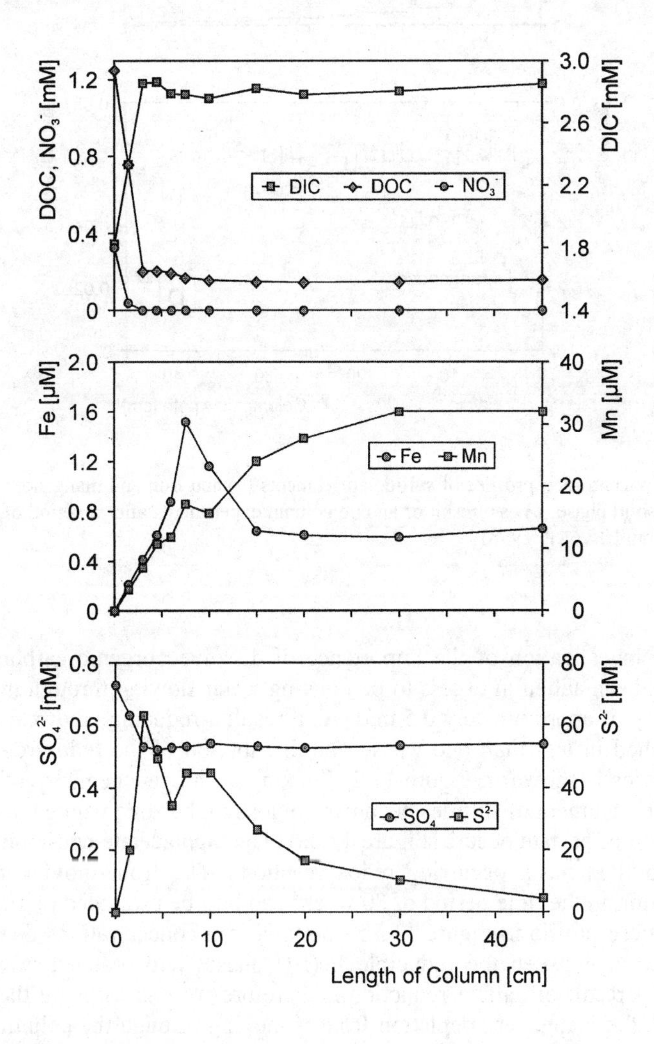

Fig. 11.2: Distribution of dissolved redox sensitive species in a laboratory column test. Sodium acetate has been added to induce redox processes at a flow velocity of approximately 0.5 m/d (modified from HENCKE, 1998).

by a very low organic carbon content. The experiments were run at variable concentrations of dissolved organic matter as well as different flow velocities.

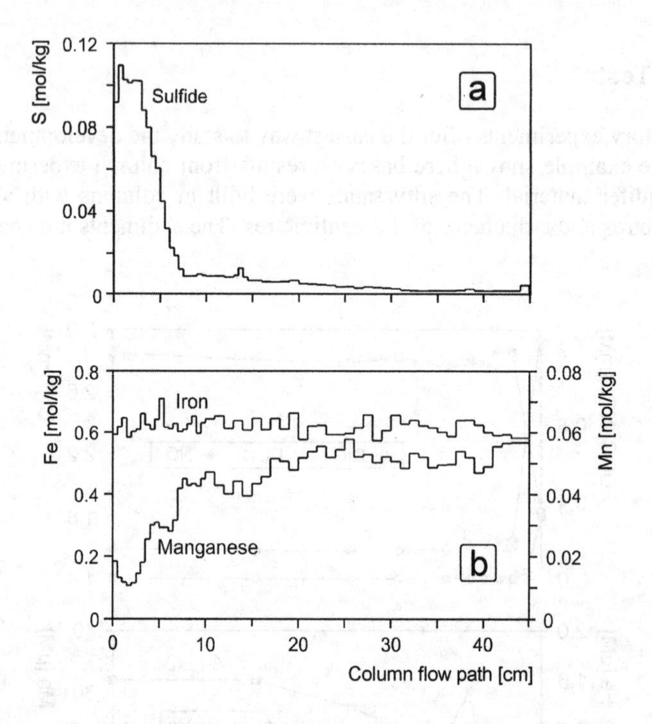

Fig. 11.3: Concentration profiles of sulfide enrichments (a) and iron and manganese concentrations in the solid phase (b) as a result of anoxic column experiments after a period of 14 months (modified from HENCKE, 1998)

For the investigation of the importance of dissolved organic carbon, sodium acetate has been added in excess to percolating water flowing through the column at a velocity of approximately 0.5 m/d. As a result a reducing environment could be established in less than one week. The distribution of the redox-relevant dissolved species is shown in Figure 11.2. Looking at the geochemistry of the solid phase, the enrichment of sulfides becomes obvious in the zone where the oxidative consumption of acetate occurs (Figure 11.3a). The high acetate consumption rates lead to simultaneously occurring redox reactions. The front moving along the flow-path during the time period of 80 weeks can best be estimated on the basis of the manganese profile in Figure 11.3b (the total iron concentrations do not allow the distinction between the reducible Fe(III) phases and reduced iron sulfides formed as a result of sulfate reduction). Therefore we can estimate the velocity with which the manganese depletion front is moving through the column to be in the range of a few centimetres per year.

11.3.1.1
Flow Rate Versus Decomposition Rate

At this stage it is important to discuss the influence of the flow velocity on the organic carbon decomposition rate. The transport of dissolved species may be due to advection or diffusion. In marine or limnic sediments concentration profiles in the pore-water are predominantly controlled by diffusive processes, whereas in most aquifers hydraulic gradients contribute to an advective groundwater flow. ROWE (1987) has found that the hydromechanical dispersion becomes insignificant with respect to molecular diffusion at transport velocities less than $3 \cdot 10^{-9}$ m/s or approximately 9 cm/a. At higher groundwater flow velocities the question arises which effect does advection have on the front movement. Basically there are two scenarios possible:

1. The reaction rate remains constant leading to increasing supplies of reactive species at increasing flow velocities.
2. The reaction rate increases due to the increasing supply of oxidising agents leading to a higher bacterial population density. In addition, microbial activity may be enhanced by fast removal of growth-inhibiting metabolic products.

In the latter case, redox fronts may even migrate against the flow direction when groundwater velocity is high. Investigations by HUETTEL et al. (1998) have revealed that the microbial reaction rate in marine sediments strongly increases once a lateral flow at the sediment/water interface induces advective pore-water flow within the upper centimetres of the sediment.

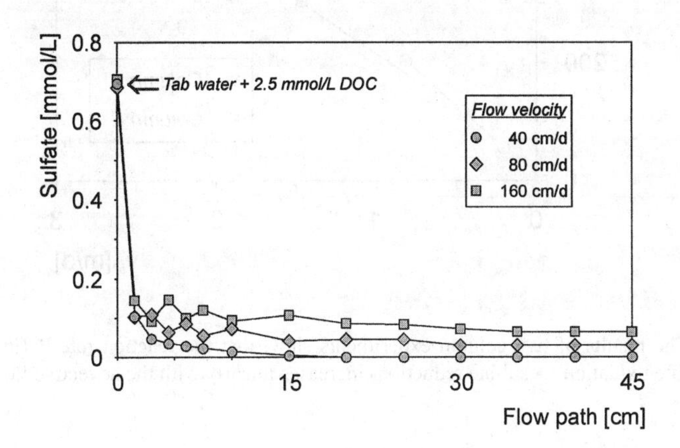

Fig. 11.4: The decomposition rate of acetate added to the water increases due to a stepwise rise of the flow velocity.

Figure 11.4 shows the results of column experiments where the flow velocity has been varied, whereas the concentration of dissolved organic substance (sodium acetate) remains constant. Although the flow velocity has been increased stepwise from 40 to 160 cm/d, it is obvious that the location of maximum reaction rates remains within the first centimetres of the flow-path. In Figure 11.5 the sulfate reduction rates R are plotted against the advective flow v_a and it becomes obvious that R increases almost linearly. However, Figure 11.4 shows that sulfate reduction is complete after some 15 cm only at the lowest flow velocity of approximately 40 cm/d. At higher velocities the sulfate reduction is incomplete, leading to constant concentrations of 0.05 and 0.1 mmol/L, respectively. Since the flow velocity was increased stepwise from low to high velocity the microbiological population density was probably very low further down the flow-path. All electron donors have been reduced along the first centimetres of the column leading to environmental conditions in which bacterial life became very difficult. With increasing flow velocity, the decomposition rate of sulfate reached a level where the residence time of the sulfate-bearing water in the zone of maximum reduction rates is too short for complete reduction. Due to the high flow velocity some of the remaining sulfate was transported into areas in which microbial population is very sparse, leading to low rates of sulfate reduction.

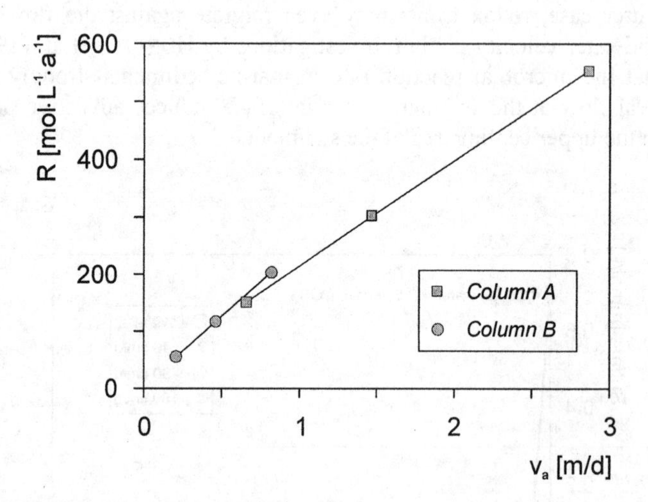

Fig. 11.5: The results of two column experiments show that the reaction rate R (in this case sodium acetate oxidation ↔ sulfate reduction) increases linearly with the advective flow v_a.

It is obvious that these effects depend on the bio-availability of the organic substance. Due to these effects there is a range of decomposition rates which may buffer varying transport velocities. This buffer decreases with the bio-availability and is very small in natural systems were the decomposition of organic matter in the sediments prevails.

11.3.2
Lake Sediments

In natural systems the distribution of organic substance is generally much more heterogeneous. In most cases the reducing capacity of aquifer systems is due to the low reactivity of organic substances which have been deposited at the same time as the aquifer sediments. As a result, most of the bio-available fractions have been decomposed in the course of geological time leading to the low reducing capacities described above. The importance of "reactive" organic matter for the geochemical development along a flow-path may be illustrated in an exfiltrating lake, where constant deposition of organic-rich sediments leads to the establishment of redox fronts.

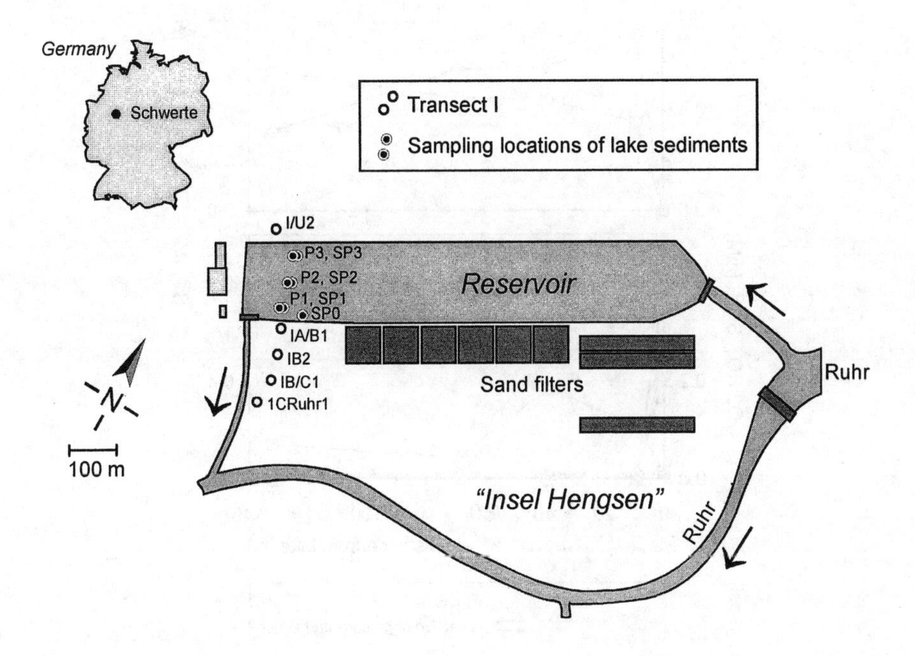

Fig. 11.6: Regional setting of the water works "Insel Hengsen" (modified from HENCKE, 1998.)

Figure 11.6 shows the regional setting at the water works "Insel Hengsen" near Schwerte, Germany. For the infiltration of surface water the Ruhr river (mean discharge of 15 m^3/s) is branched to flow through the reservoir where approximately 0.15 m^3/s are pumped off and infiltrated through sand filters (SCHULTE-EBBERT et al., 1991).

At the south-western end of the reservoir a transect of observation wells has been built, which is hydraulically not influenced by the infiltration of water through the sand filters in the east of the plant. Therefore, the hydrochemical development during bank infiltration can be studied. The underlying Quaternary aquifer has an average thickness of 5 metres and consists of alluvial sediments with hydraulic conductivities in the range of $k_f = 10^{-2}$ and 10^{-3} m/s. It is covered by up to 2 metres of alluvial sediments. The aquifer base is formed by carboniferous silty clays of the Carboniferous (SCHÖTTLER & SOMMER, 1987). The reservoir covers an area of some 20 hectares and has an average depth of 2 metres. The greatest thickness of the autochthonous lake sediments is found in the central area where they may be thicker than 2 metres. Considering an operation period of 60 years, this corresponds to an annual sedimentation rate of approximately 3.5 cm/a.

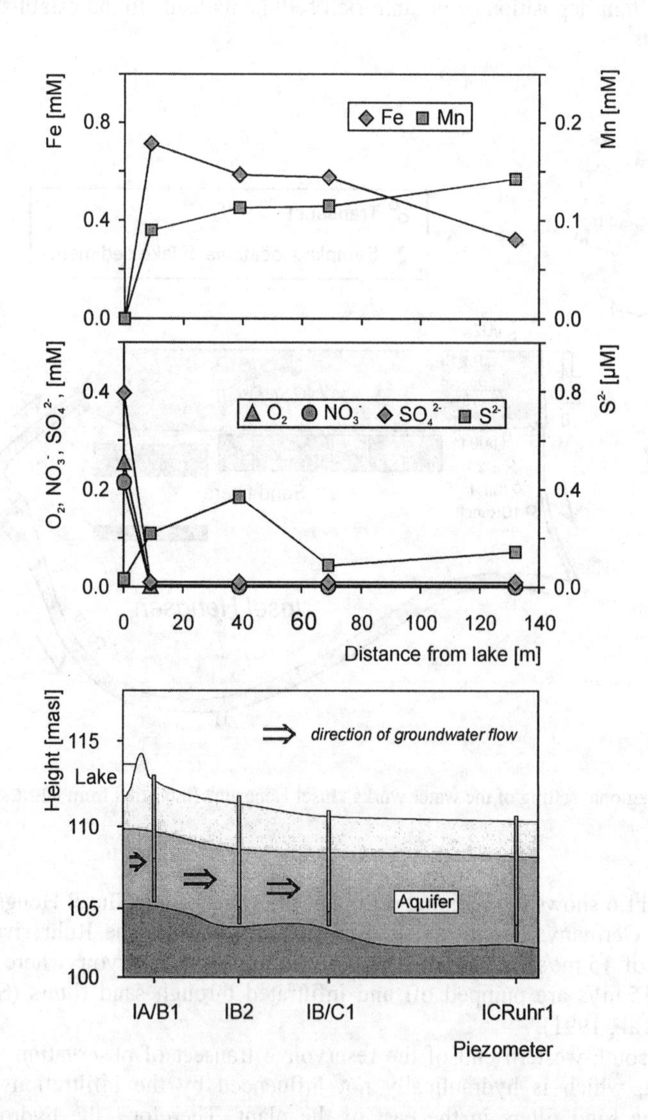

Fig. 11.7: Schematic overview of the redox driven hydrochemical development of lake water infiltrating into the aquifer along Transect I (modified from HENCKE, 1998).

The geochemical conditions in the transect are characterised by redox processes leading to complete sulfate reduction. Iron concentrations approach 1 mmol/L due to reductive dissolution of ferric iron minerals (Figure 11.7). Balancing the electron transfers for the oxidation of dissolved organic carbon (DOC) revealed a lack of electron donors corresponding to 20 mg/L DOC (LENSING, 1995; SCHULTE-EBBERT & SCHÖTTLER, 1995). The concentration of organic carbon contained in the aquifer sediments is extremely low (0.2 %wt) (SELENKA & HACK, 1992), only the lake sediments may serve as a source for organic carbon (8.0 %wt). Several Darcy experiments have revealed a uniform hydraulic conductivity of $k_f = 3 \cdot 10^{-6}$ m/s for a profile extending from the transect through the western part of the lake. On the basis of a hydraulic model of the profile WENZEL (1998) could show that a flow velocity of approximately 0.8 m/d is possible within the lake sediments due to the infiltration of surface water.

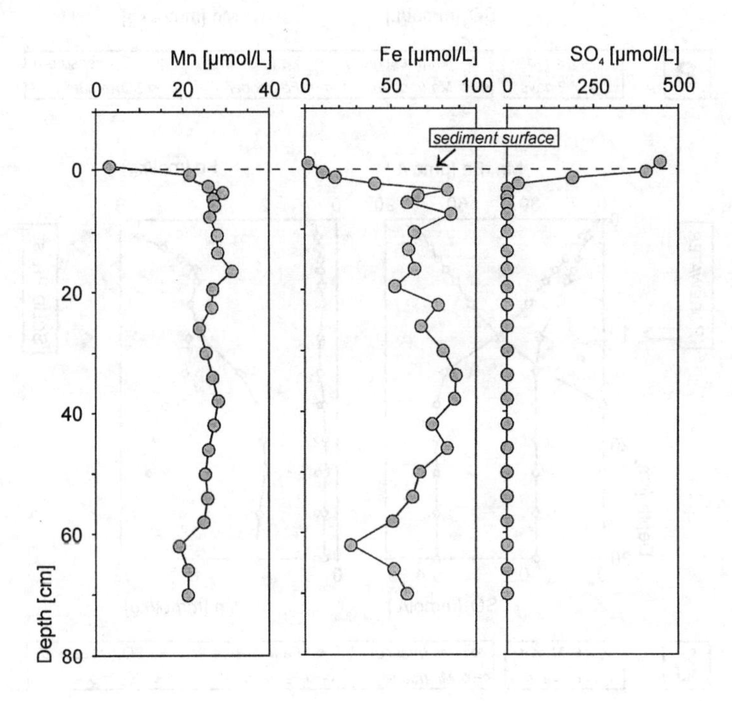

Fig. 11.8: Porewater profiles of iron and sulfate within the lake sediments at the location SP0 shown in Figure 11.6 (modified from CHRISTENSEN, 1997).

Figure 11.8 shows pore-water profiles within in a typical core of the lake sediments. This particular core has been retrieved from an area close to the southern edge of the lake. Advective flow velocities are very low and diffusive processes prevail at this site. However, the sulfate concentrations in the cores retrieved along the transect in the lake showed greater penetration depths approaching 17 cm with

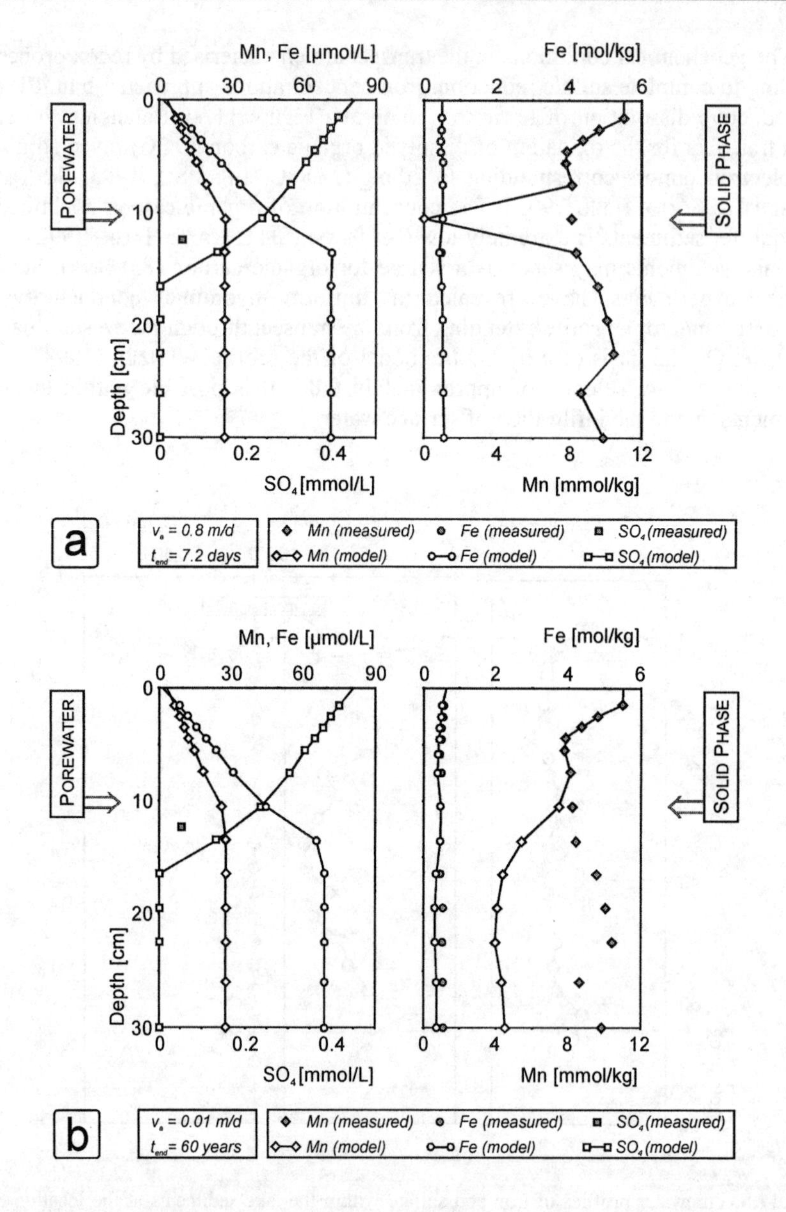

Fig. 11.9: The coupled modelling of transport and reaction reveals that reducible manganese minerals would be depleted within the lake sediments presuming a flow velocity of 0.8 m/d resulting from hydraulic modelling (a). At a transport velocity in the order of 1 cm/d an equilibrium between reduction rate and sedimentation rate is possible (b).

respect to sulfate. It can be assumed that the bio-availability of the organic substance within the lake sediments is regionally constant and therefore the transport velocity must be higher in the central regions of the lake. It is obvious that the geochemical processes leading to the groundwater composition observed in the first well of Transect I almost completely take place within the lake sediments. It could be shown by the coupled modelling of transport and reaction with the model CoTReM (LANDENBERGER, 1998) that manganese minerals are quickly depleted in the lake sediments due to reductive solution at flow velocities of 0.8 m/d (Figure 11.9a). Only at flow velocities of approximately 0.01 m/d is a steady-state established between the sedimentation rate of 3.5 cm/a and the reductive solution of manganese minerals (Figure 11.9b).

From the observations along Transect I it is obvious that there is a sufficient amount of reducible manganese minerals within the sediments. Taking into account that the lake sediments have reached their thickness over a period of 60 years it becomes clear that manganese reduction rates have been low enough at all times to prevent complete depletion of manganese minerals. This rate is determined by the depth of manganese reduction, the flow velocity and the dissolved manganese concentrations in Transect I. Due to constant sedimentation of oxidised manganese minerals, the steady-state reaction front is moving upwards within the sediment, relative to the velocity of sedimentation which is in the range of a few centimetres per year.

11.3.3 Bank Infiltration

As we have seen, the abundance and bio-availability of organic substance is crucial for the formation and the lateral scale of redox fronts. In many cases the amount of bio-available organic matter is insufficient to reduce all available electron donors. In rare cases the low reactivity may lead to a redox succession extending over more than 100 kilometres as seen in Figure 11.1.

The Oderbruch in the eastern state of Brandenburg, Germany, covers an area of some 80,000 hectares and represents the largest enclosed river polder in Germany (see Figure 10.13). The polder is bordered by the river Oder in the east and by Geest high-plains in the west. The Oder river was re-directed 250 years ago in order to reclaim agricultural land. As a result, the river embankment is situated above the groundwater table within the polder, hence river water infiltrates into the polder aquifer. Along this infiltration a number of piezometer transects have been installed, one of which is shown in Figure 11.10. The aquifer consists of an upper and a lower system. Looking at the concentrations of redox sensitive compounds it is obvious that the upper system has a reduction capacity which leads to the complete reduction of sulfate, whereas sulfate reduction is not significant in the lower system. This is due to the different abundance of reactive organic matter. Whereas the lower system consists of glacial sands, the upper system is built up of post-glacial alluvial sediments containing higher amounts of organic carbon this is also illustrated in considerably higher concentrations of dissolved organic carbon in the pore-water. This fact is supported by $\delta^{34}S$ measurements in both systems. Starting from a $\delta^{34}S$-value of +5.0 ‰ in the Oder river, a high accumulation of the heavy isotope is observed in the shallow aquifer ($\delta^{34}S = +85.7$ ‰), where dissolved sulfate concentrations reach a minimum at piezometer 9534 (Figure 11.10). Further along the flow-path sulfate concentrations increase again. This

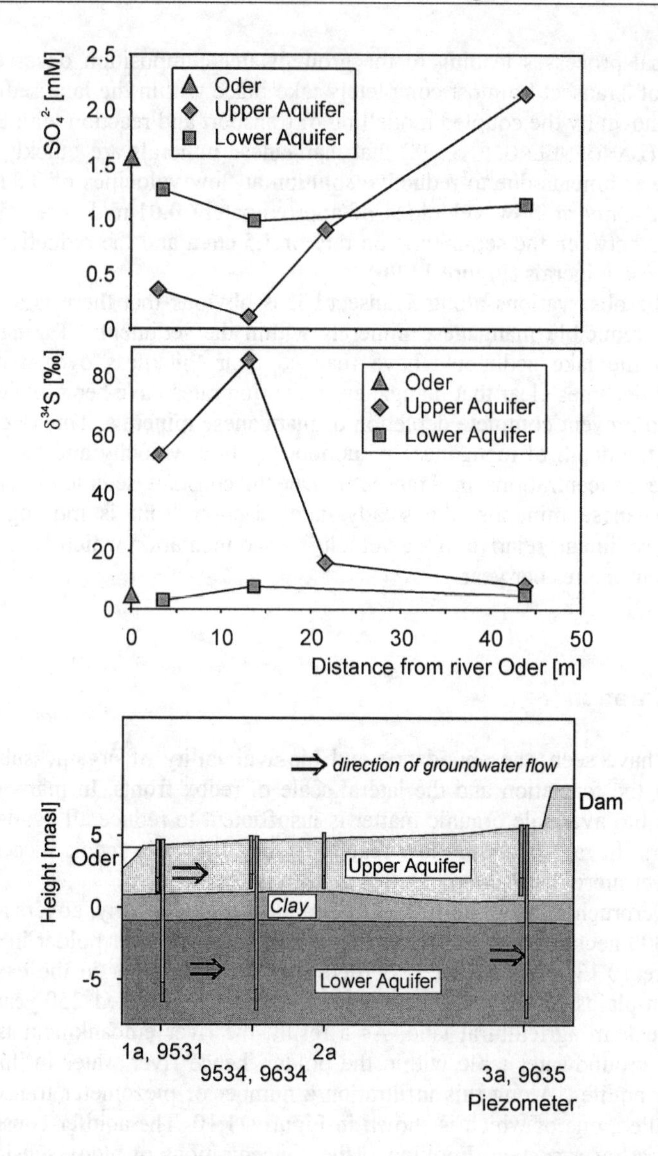

Fig. 11.10: Concentrations of redox-sensitive parameters along the infiltration flow-path of transect "Bahnbrücke" in the north-eastern Oderbruch polder.

is due to the fact that the river bank is flooded regularly towards the dam. Under water-logged conditions a reducing environment is established and sulfides may be formed in the soils which are exposed to atmospheric oxygen at times of normal water levels. The sulfide oxidation is then illustrated by rising sulfate concentrations and decreasing δ^{34}S values.

In comparison, the increase of δ^{34}S-values in the lower system to +7.9 ‰ reflects the minor importance of sulfate reduction. Again we can conclude that the amount of reactive organic carbon decides whether a complete redox sequence can be observed or not. The amount of organic carbon is just sufficient for some iron reduction. The amount of reducible iron compounds in the aquifer then controls a redox front moving along the flow-path, leaving behind an "iron-free" aquifer. The continuing supply of organic carbon dissolved in the infiltrating water from the Oder river would then be available for sulfate reduction. A very rough estimate of the front velocity can be made on the basis of the hydraulic conditions and the chemical composition with respect to dissolved organic carbon and ferric oxide pools. A rough estimation of an iron depletion velocity can be done by means of computer models such as CoTReM coupling transport and reaction processes which allow the combined modelling of one-dimensional transport and reaction (Landenberger, 1998). Assuming a groundwater velocity of approximately 25 cm/d through the aquifer sediments containing approximately 1 g/kg reducible Fe(III), a reaction rate of $3.3\cdot10^{-4}$ $mol\cdot l^{-1}\cdot a^{-1}$ would lead to dissolved iron concentrations in the range of those observed along the transect. Therefore, after 1000 years the iron depletion front would be in the order of 10 to 100 metres away from the River Oder.

11.4 Summary

The establishment of reaction fronts driven by redox processes is linked to the abundance and reactivity of bio-available organic matter, the amount of redox-sensitive components in the solid phase and the transport processes. In case of organic substances bound to the solid phase, the bio-available fraction will be depleted over time leading to a redox-front moving along the flow-path. Due to gradually decreasing gradients the velocity with which the front moves will slow down. If the organic substance is dissolved, the redox-cline will establish at a certain distance which depends on the equilibrium between decomposition rate and the flux rate. As shown in the column experiments, the micro-organisms may adjust their activity over a wide range. This again depends on the flow velocity as well as the bio-availability of the organic substance which is high in the case of sodium acetate.

For the movement of a steady-state front in aquifers driven by redox reactions one has to distinguish between dissolved organic carbon and organic substances bound to the aquifer material. In the latter case, the front moves along the flow-path together with the depletion of reactive organic carbon. A second front may be established by the depletion of reducible manganese and iron minerals. In the case of a continuous supply of dissolved organic carbon, the position of the redox front depends on the microbial reaction rate and the velocity with which reactants are delivered. The lateral movement would then be synchronous with the depletion of reducible minerals in the aquifer.

The maximal decomposition rate of organic matter depends on the bio-availability – or reactivity – of the organic substance. Whereas dissolved acetate is a highly reactive substrate for bacteria mediating redox processes, many organic substances bound to the solid phase are refractive upon. Modelling the reactive transport we may think of the decomposition rate to be highly variable depending on the reactivity of the organic substance. This is illustrated by the column tests which the transport velocity may be tripled without a noteworthy lateral shift of the site of decomposition. In natural aquifer systems, redox conditions are often characterised by low reduction capacities. This is due to the low reactivity of geogenic organic substances and the limited input of dissolved organic substances.

11.5 References

BERNER, R.A. (1981): Early diagenesis: A theoretical approach. Princeton University Press, 241pp.

BRADLEY, P.M.; FERNANDEZ, JR., M. & CHAPELLE, F.H. (1992): Carbon limitation of denitrification rates in an anaerobic groundwater system. Environ. Sci. Technol. 26: 2377-2381.

CALVERT, S.E. & PEDERSEN, T.F. (1993): Geochemistry of recent oxic and anoxic marine sediments: Implications for the geological record. Mar. Geol. 113: 67-88.

CHRISTENSEN, B. (1997): Geochemische Charakterisierung eines Sedimentkernes aus dem Stausee Hengsen (Schwerte). unveröff. Diplomarbeit, Fachbereich Geowissenschaften, Universität Bremen, 85pp.

CORD-RUWISCH, R.; SEITZ, H.-J. & CONRAD, R. (1988): The capacity of hydrogentropohic anaerobic bacteria to compete for traces of hydrogen depends on the redox potential of the terminal electron acceptor. Arch. Microbiol. 149: 350-357.

FRANCOIS, R. (1988): A study on the regulation of the concentration of some trace metals (Rb, Sr, Zn, Pb, Cu, V, Cr, Ni, Mn and Mo) in Saanich Inlet sediments, British Columbia, Canada. Mar. Geol. 83: 285-308.

HENCKE, J. (1998): Redoxreaktionen im Grundwasser: Etablierung und Verlagerung von Reaktionsfronten und ihre Bedeutung für die Spurenelement-Mobilität. Ber. Fachber. Geowissenschaften Univ. Bremen 128: 122pp.

HUETTEL, M.; ZIEBIS, W.; FORSTER, S.; LUTHER III, G.W. (1998): Advective transport affecting metal and nutrient distributions and interfacial fluxes in permeable sediments. Geochim. Cosmochim. Acta 62: 613-631.

KÖLLE, W.; WERNER, P.; STREBEL, O. & BÖTTCHER, J. (1983): Denitrifikation in einem reduzierenden Grundwasserleiter. Vom Wasser 61: 125-147.

LANDENBERGER, H. (1998): CoTReM, ein Multi-Komponenten Transport- und Reaktions-Modell. Ber. Fachber. Geowissenschaften Univ. Bremen 110: 142pp.

LENSING, H.J. (1995): Numerische Modellierung mikrobieller Abbauprozesse im Grundwasser. Mitt. Inst. Hydrologie und Wasserwirtschaft Univ. Karlsruhe. 51: 185pp.

LOVLEY, D.R. & GOODWIN, S. (1988): Hydrogen concentrations as an indicator of the predominant terminal electron-accepting reactions in aquatic sediments. Geochim. Cosmochim. Acta 52, 2993-3003.

LOVLEY, D.R. & PHILLIPS, E.J.P. (1987): Competitive mechanisms for the inhibition of sulfate reduction and methane production in the zone of ferric iron reduction in sediments. Appl. Environ. Microbiol. 53, 2636-2641.

POSTMA, D.; BOESEN, C.; KRISTIANSEN, H. & LARSEN, F. (1991): Nitrate reduction in an unconfined sandy aquifer: water chemistry, reduction processes and geochemical modeling. Water Resour. Res. 27, 2027-2045.

POSTMA, D. & JAKOBSEN, R. (1996): Redox zonation: Equilibrium constraints on the Fe(III)/SO_4-reduction interface. Geochim. Cosmochim. Acta 60, 3169-3175.

ROWE, R.K. (1987): Pollutant transport through barriers. Proc. Geotech. Practice for Waste Disposal '87, 159-181.

SCHÖTTLER, U. & SOMMER, H. (1987): Hydrogeologische Untersuchung des Ruhrtals im Bereich der Wassergewinnungsanlagen der Dortmunder Stadtwerke AG. Bericht der Dortmunder Stadtwerke AG 5.

SCHULTE-EBBERT, U.; HOLLERUNG, R.; WILLME, U.; KACZMARCZYK, B.; BAHRIG, B. & SCHÖTTLER, U. (1991): Verhalten von anorganischen Spurenstoffen bei wechselnden Redoxverhältnissen im Grundwasser. Dortmunder Beiträge zur Wasserforschung 43: 358pp.

SCHULTE-EBBERT, U. & SCHÖTTLER, U. (1995): Systemanalyse des Untersuchungsgebietes „Insel Hengsen". In: SCHÖTTLER, U. & SCHULTE-EBBERT, U. (Eds.): Schadstoffe im Grundwasser – Bd. 3: Verhalten von Schadstoffen im Untergrund bei der Infiltration von Oberflächenwasser am Beispiel des Untersuchungsgebietes „Insel Hengsen" im Ruhrtal bei Schwerte. VCH Weinheim, 475-513.

SELENKA, F. & HACK, A. (1992): Quantifizierung und Charakterisierung der organischen Substanz am Korngerüst des Bodens und der Sedimente im Versuchsfeld Hengsen. unpubl. report of the working group „Gewinnung von Festmaterial für Analysen und Laborversuche" of DFG-Priority Program „Schadstoffe im Grundwasser", DFG Bonn, 231pp.

VAN BEEK, C.G.E.M.; BOUKES, H.; VAN RIJSBERGEN, D. & STRAATMAN, R. (1988): The threat to the Netherlands waterworks by nitrate in the abstracted groundwater, as demonstrated on the well field Vierlingsbeek. Wat. Supply 6, 313-318.

VAN BEEK, C.G.E.M. & VAN DER KOOIJ, D. (1982): Sulfate-reducing bacteria in groundwater from clogging and nonclogging shallow wells in the Netherlands river region. Ground Water 20, 298-302.

WENDLAND, F.; ALBERT, H.; BACH, M. & SCHMIDT, R. (1994): Potential nitrate pollution of groundwater in Germany: A supraregional differentiated model. Envir. Geol. 24, 1-6.

Chapter 12

Redox Processes Active in Denitrification

C.G.E.M. van Beek

12.1
Introduction

Intensive agriculture in the Netherlands is characterised by the application of massive amounts of nitrogen to the fields. These heavy dressings are applied as liquid manure and nitrogen-fertiliser. The excess nitrogen leaches to the groundwater, and shows up as increasing nitrate concentrations in groundwater and consequently in surface water as well. This phenomenon is especially prominent in sandy regions. In the case of liquid manure application high concentrations of nitrate are accompanied by high concentrations of potassium.

Groundwater abstracted by well-fields from shallow phreatic aquifers also show increasing nitrate concentrations. However, many well-fields situated in these mentioned critical conditions, i.e. a shallow aquifer and intensive dressings of liquid manure and/or nitrogen fertiliser, show only a limited increase in the concentration of nitrate, or no nitrate at all. Apparently denitrification is active around these well-fields. Consequently the disappearance of nitrate should accordingly be accompanied by concentration changes of other parameters involved in the denitrification process.

Information about changes in the chemical composition of groundwater is important for those responsible for the quality of groundwater, and for water works in particular. In this chapter we want to look at the following questions:

- which compounds are active in denitrification;
- which processes govern the contribution of these compounds; and
- which other changes in the chemical composition of groundwater may be expected due to denitrification?

We will answer these questions with the help of data obtained from a multilevel miniscreen observation well, located near the well-field Vierlingsbeek.

12.2 Site and Methods

The well-field Vierlingsbeek is located in the south-eastern part of the Netherlands. The production of the well-field is about $2.8 \cdot 10^6$ m^3/a. The volume is abstracted with the help of about 10 production wells. The screens of these wells extend between about 20 to 30 meters below the surface (mbs).

The phreatic aquifer near the well-field Vierlingsbeek is composed of unconsolidated fluvial sands, containing no calcite, and low amounts of organic matter (0.1 to 2%) and pyrite (<0.01 to 0.2%). The contents of both substances increase with depth. The clay content is also very low. The geohydrological base consists of fine-grained cemented deposits at about 30 mbs. The depth of the groundwater table varies between 2 and 4 mbs.

The well-field itself is located in a small forest, which is surrounded by agriculture on all sides. In this area, very intensive cattle breeding is prominent, consequently heavy dressings of liquid manure have been applied. Recently government measures have been imposed to reduce these dressings to environmentally and agriculturally sound amounts.

The multilevel miniscreen observation wells are equipped with 20 to 30 miniscreens at regular distances of 1 m, all attached to a central PVC support pipe. A miniscreen consists of a gravel-coated slotted PVC pipe (length 6 cm, ∅ 2.5 cm). A teflon tube connects each screen to the surface and allows in this way sampling of groundwater by vacuum extraction. The extracted groundwater samples have been analysed by conventional methods.

12.3 Geochemistry

Figure 12.1 shows the results of the chemical analysis of groundwater sampled from NP40, from 6 down to 28 mbs. Presented is the relation between depth and respectively the pH and the concentrations of redox-sensitive species: nitrate, sulfate, iron, manganese, arsenic, total inorganic carbon (TIC), dissolved organic carbon (DOC) and redox non-sensitive species: potassium, sodium, hardness and chloride. High concentrations of nitrate and potassium are indicative of heavy

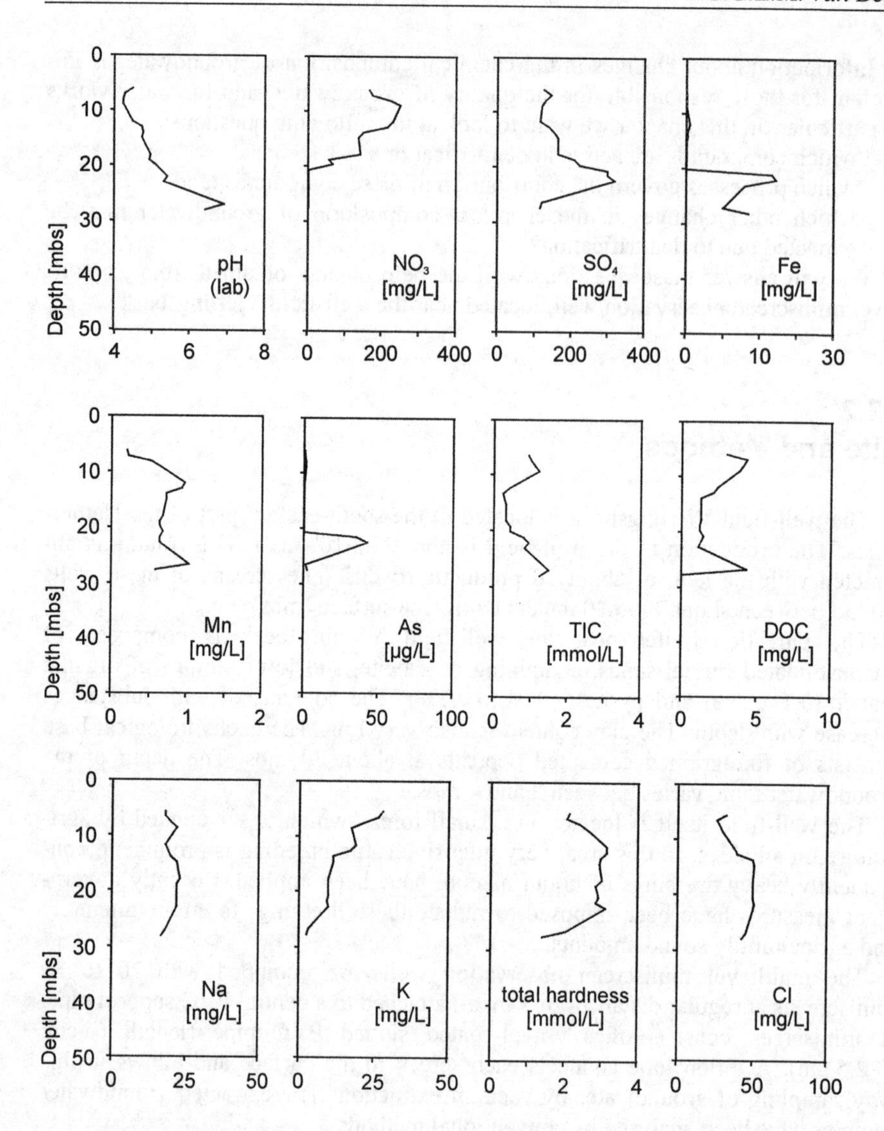

Fig. 12.1: Depth profiles of pH, total hardness (calcium and magnesium) and concentrations of sodium, potassium, nitrate, sulfate, iron, manganese, arsenic, chloride, Total Inorganic Carbon (TIC) and Dissolved Organic Carbon (DOC), of NP40.

dressings of liquid manure. The excess applied is leached to the groundwater. Natural (pristine) background concentrations of nitrate and potassium in groundwater do not exceed 1 to 2 mg/L. Figure 12.1 shows that the concentration of potassium at 23 meters below surface (mbs) equals about 10 mg/l. The concentration of potassium should be in equilibrium with the amount adsorbed by the cation-exchange complex. The increased concentration of potassium will result in an adsorption of potassium. Even in the case of minute amounts of clay, the "potassium front" by definition lags behind the "manure front", which must have been penetrated deeper than 23 mbs.

Figure 12.1 also demonstrates that the groundwater below 21 mbs does not contain nitrate, which means the "nitrate front" is less deep than the "potassium front". Nitrate can only disappear from groundwater by denitrification. As the "nitrate front" has penetrated less than the "potassium front", it may be deduced that considerable denitrification has occurred.

12.3.1 Denitrification

Figure 12.1 demonstrates that the decrease of the nitrate concentration coincides with an increase of sulfate. Besides the changes in the concentration of nitrate and sulfate, changes in concentration of other parameters are negligible between 13 and 21 mbs. This points to a uniform source of the groundwater in this depth interval.

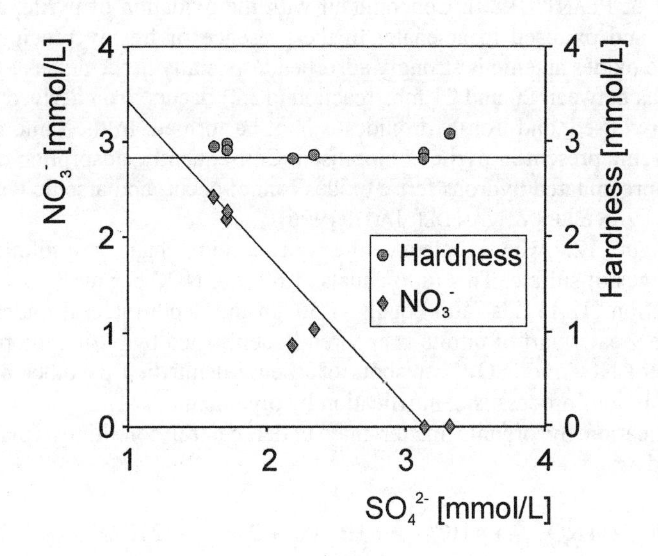

Fig. 12.2: Relationship between the concentrations of sulfate and nitrate in NP40 over the depth interval between 13 and 21 mbs.

Figure 12.2 shows the relation between the concentration of sulfate and the concentration of nitrate in this interval between 13 and 21 mbs. Between the concentration of both parameters a linear relationship seems to exist.

Figure 12.2 also shows the relation between the concentration of sulfate and the total hardness. It remains constant, which corroborates the assumption of a the homogeneous depth-interval, and also shows that calcium and magnesium are not involved in any of the reactions.

The negative linear relation between nitrate and sulfate may be most conveniently explained by denitrification caused by the oxidation of sulfides, i.e. pyrite. In the presence of excessive nitrate, denitrification by pyrite may be represented as:

$$2\,FeS_{2(s)} + 6\,NO_3^- + 4\,H_2O \rightarrow 2\,Fe(OH)_{3(s)} + 4\,SO_4^{2-} + 3\,N_{2(g)} + 2\,H^+ \tag{12.1}$$

and in the case of an excess pyrite by:

$$2\,FeS_{2(s)} + 14\,NO_3^- + 4\,H^+ \rightarrow 5\,Fe^{2+} + 10\,SO_4^{2-} + 7\,N_{2(g)} \tag{12.2}$$

It can be clearly seen from Figure 12.1 that no iron is present in solution between 13 and 20 mbs. This implies that the process is governed by reaction (12.1). The occurrence of reaction (12.1) (excess nitrate versus excess pyrite) is corroborated by the behaviour of arsenic. Pyrite is known to contain traces of arsenic (RAISWELL & PLANT, 1980). Concomitant with the oxidation of pyrite, arsenic is mobilised, and oxidised to arsenate. In the presence of freshly precipitated hydrous ferric oxides arsenic is strongly adsorbed, especially in the arsenate form.

At depths between 20 and 21 mbs reaction (12.2) occurs. Iron derived from pyrite is not oxidised, and iron-hydroxides cannot be formed. In the same oxidation process arsenic present in pyrite is mobilised. Consequently, adsorption of arsenic by freshly precipitated hydrous ferric oxides cannot occur, and arsenic will remain in solution (VAN BEEK & VAN DER JAGT, 1996).

From Figure 12.2 we can calculate the concentration changes of nitrate relative to the changes of sulfate. This ratio equals -1.60 mol NO_3^- per mol SO_4^{2-} According to reaction (12.1) this ratio equals -1.50. In the depth-interval under consideration, the greater part of nitrate is apparently denitrified by pyrite, the remaining part, 0.1 mol NO_3^-/mol SO_4^{2-}, or about 6% being denitrified by other processes. The most obvious process is denitrification by organic matter.

Denitrification by organic matter may under neutral conditions (pH>6.4) be represented as:

$$5\,CH_2O_{(s)} + 4\,NO_3^- \rightarrow 4\,HCO_3^- + H_2CO_3 + 2\,N_{2(g)} + 2\,H_2O \tag{12.3}$$

and under acid conditions (pH<6.4) as:

$$5\,CH_2O_{(s)} + 4\,H^+ + 4\,NO_3^- \rightarrow 5\,H_2CO_3 + 2\,N_{2(g)} + 2\,H_2O \tag{12.4}$$

Figure 12.1 indeed shows a small increase in the concentration of TIC. However if we assume that 6% of 2.4 mmol/l NO_3^- is at maximum reduced by organic matter, this will result in an increase in the concentration of TIC of less than 0.2 mmol/l. This increase is about equal to the increase between 13 and 20 mbs.

In reaction (12.1) 0.33 protons per mole nitrate denitrified are produced, whereas one proton is consumed for each mole nitrate denitrified in reaction (12.4). Taking the observed distribution of both denitrification processes into account, this should result into a net proton production. However, from Figure 12.1 it is clearly seen that the pH increases with depth, indicating an overall proton consumption. Apparently, more acid-base processes are active. POSTMA et al. (1991) found the same phenomenon.

Between 20 and 21 mbs reaction (12.2) occurs. In this reaction protons are consumed, which is corroborated by a rise in pH. Moreover, the concentration of TIC increases, which indicates the contribution of organic matter.

At greater depth, sulfate reduction apparently does occur, but this process is not considered here.

12.3.2 Reduction Capacity

Table 12.1 shows the half reactions of both reductants, the reduction capacity of organic carbon amounting to 4 electron equivalents (e^-) per mole organic carbon, or 0.33 e^-/g organic carbon. As the carbon content of organic matter equals about 0.57, this means that the reduction capacity of organic matter equals 0.19 e^-/g organic matter. The reduction capacity of pyrite equals 15 e^-/mol FeS_2, or 0.14 e^-/g FeS_2. Per unit mass, the order of magnitude for both reductants is thus comparable.

Tab. 12.1: Electron equivalents for solid redox compounds.

Half Reaction	*e^- Equivalents* *
$CH_2O + H_2O \rightarrow CO_2 + 4H^+ + 4e^-$	$4\{CH_2O\}$
$FeS_2 + 11H_2O \rightarrow Fe(OH)_{3(s)} + 2SO_4^{2-} + 19H^+ + 15e^-$	$15\{FeS_2\}$

* Electron equivalents (e^-) are equal to the molar content multiplied by the absolute number of electrons transferred.

Figure 12.3 shows the content of organic matter and pyrite as a function of depth of multi-level miniscreen observation well NP1, which is located near by NP40. The contents of pyrite varies between <0.01 and 0.2% (w/w), and for organic matter between 0.1 and 2% (w/w).

Figure 12.3 demonstrates that the weight percentage of pyrite is about a factor of 10 lower than the content of organic matter. The reduction capacity of pyrite and organic matter being the same order of magnitude shows that in NP1 the reduction capacity of organic matter per unit weight of soil is by far greater than that of pyrite. Yet, denitrification by organic matter is thermodynamically favourable

(KOROM, 1991) and the excess of organic matter over pyrite, nitrate nevertheless is reduced by pyrite. This means that the reactivity of organic matter in denitrification is much lower than the reactivity of pyrite (BÖTTCHER al., 1991).

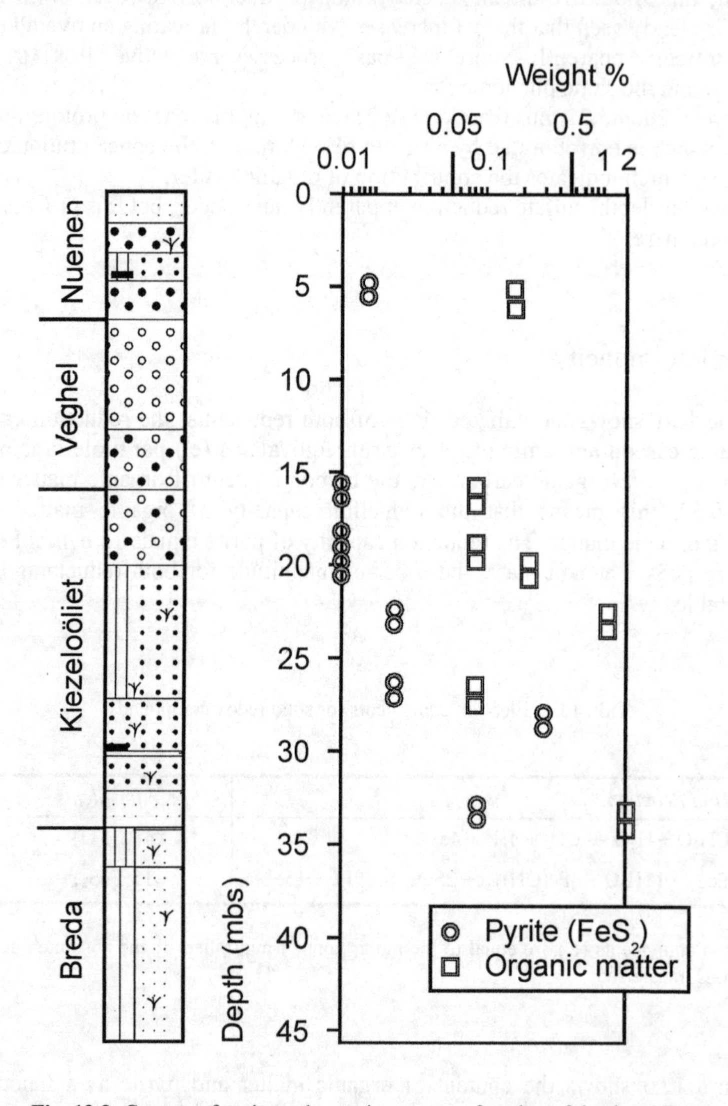

Fig. 12.3: Content of pyrite and organic matter as function of depth at NP1.

The continuous and depth-dependant decrease in the concentration of nitrate in the presence of pyrite is very conspicuous. The greater part of the aquifer contains pyrite in quantities below the analytical detection limit, i.e. 0.01% (w/w). This does not necessarily imply that pyrite is not present. However, present in very low percentages (<0.01%), micro-morphological research has confirmed the presence of pyrite in nitrate containing groundwater. BISDOM & BREEUWSMA (1990) reported the presence of pyrite in the aquifer near Vierlingsbeek in the form of framboids, which are present inside plant detritus. In such a way, pyrite is apparently shielded from an oxidising environment.

Due to this occurrence, apparently, the sulfides become gradually available in denitrification, and in this way also explain the gradual disappearance of nitrate with depth. Actually, in the case of an abundant pyrite supply, one should expect a sharp drop in nitrate concentration.

12.4 Conclusions

Based upon this research we draw the following conclusions:

1. Secondary pyrite, naturally occurring in the aquifer, plays a dominant role in denitrification.
2. Even traces of pyrite below the analytical detection limit of 0.01 % (w/w) are important.
3. In the aquifer the reduction capacity of organic matter is much greater than that of pyrite, however, organic matter plays only a minor role in denitrification. Apparently, the reactivity of organic matter is very low.
4. It is important to distinguish between two modes of denitrification, denitrification in an excess of nitrate (no iron in solution) and in an excess of pyrite (iron in solution).
5. The way in which pyrites are present determines the way in which this reaction occurs. In the case of a slow release of pyrite, nitrate is present in excess and iron is also oxidised. In the case of an excess of pyrite iron will remain in solution.
6. This behaviour also determines to a large extent the behaviour of arsenic, present in pyrite. Under mild reducing conditions (as long as nitrate is present) fresh precipitated hydrous ferric oxides are present, and arsenic is strongly adsorbed. Under strong reducing conditions (pyrite is present in excess) no freshly hydrous ferric oxides are formed. Under these conditions arsenic will remain in solution.

12.5 Acknowledgement

The co-operation with the Water Supply Company "Oost Brabant" in this research was highly appreciated.

12.6 References

VAN BEEK, C.G.E.M. & VAN DER JAGT (1996): Mobilization and speciation of trace elements in groundwater, IWSA International Workshop "Natural origin inorganic micropollutants: arsenic and other constituents", Vienna May 6-7, London.

BISDOM, E.B.A. & BREEUWSMA, A. (1990): Pyrite in sediments near pumping station Vierlingsbeek, a micromorphological and geochemical research, Staring Centrum, rapport 56, Wageningen, 53 pp. (in Dutch).

BÖTTCHER, J.; STREBEL, O.; DUYNISVELD, W.H.M. & FRIND, E.O. (1991): Reply on comment of S.F. Korom on "Modeling of multicomponent transport with microbial transformation in groundwater: the Fuhrberg case". Water Res. Res. 27: 3275-3278.

KOROM, S.F. (1991): Comment on "Modeling of multicomponent transport with microbial transformation in groundwater: the Fuhrberg case". Water Res. Res. 27: 3271-3274.

POSTMA, D.; BOESEN, C.; KRISTIANSEN, H. & LARSEN, F. (1991): Nitrate reduction in an unconfined sandy aquifer: water chemistry, reduction processes, and geochemical modeling. Water Res. Res. 27: 2027-2045.

RAISWELL, R. & PLANT, J. (1980): The incorporation of trace elements into pyrite during diagenesis of black shales, Yorkshire, England. Ec. Geol. 75: 684-699.

Chapter 13

Measurement of Redox Potentials at the Test Site "Insel Hengsen"

U. Schulte-Ebbert & T. Hofmann

13.1 Introduction

Over the past years the test site *Insel Hengsen* was in the centre of a series of research projects with divers topics, carried out by the Institute for Water Research (IfW) in co-operation with different universities and research institutes (SCHÖTTLER & SCHULTE-EBBERT, 1995; HENCKE & SCHULZ, 1998; SCHULTE-EBBERT & SCHÖTTLER, 1998). All these research projects had in common that they conducted measurements of redox potentials in groundwater systems and produced a wide range of redox potential values in this procedure, demonstrating strictly oxic to strictly anoxic, sulfate free redox conditions. A broad data base of E_H-values was collected and now gives the opportunity to have a critical look at the technical boundary conditions of the measurements carried out.

13.2 The Test Site

The test site *Insel Hengsen* is a water catchment area of the Dortmund water works (*Dortmunder Energie- und Wasserversorgung GmbH*). *Insel Hengsen* is located in the River Ruhr Valley near Dortmund, Germany (Figure 13.1). Here, the Dortmund water works ply artificial groundwater recharge by slow-sand filtration. The river water is pre-filtered by fine-gravel filters and filtrated through three slow-sand filters (each 250 x 20 m), filled with 0.7 m of fine sand. The infiltration takes place directly into the quaternary coarse-gravel of the Ruhr Valley, which lies upon non-permeable rock of the Upper Carbon facies with a thickness of about 6 m. The permeability of the gravel depends on the local contents of clay and sand and ranges from $5 \cdot 10^{-1}$ to $5 \cdot 10^{-4}$ m/s.

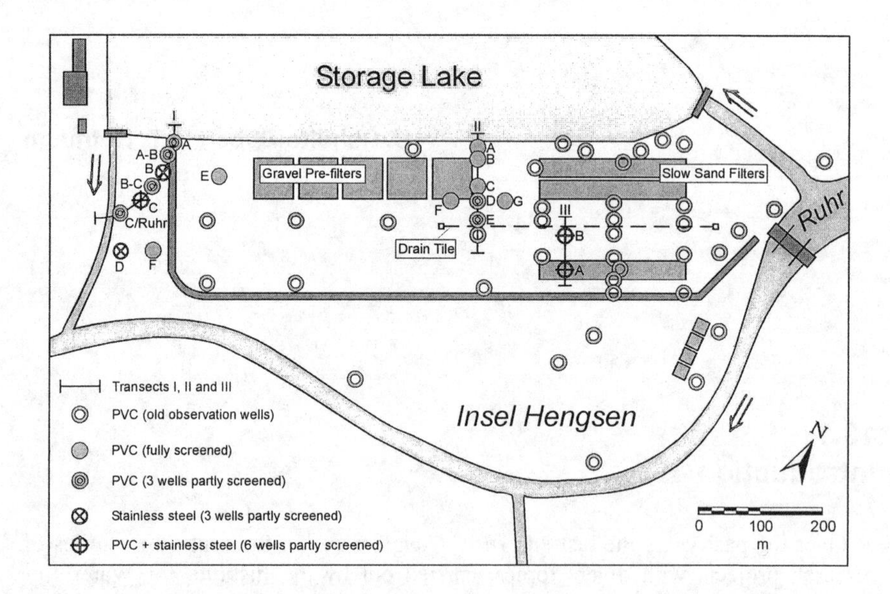

Fig. 13.1: The test site *Insel Hengsen.*

A well-equipped monitoring net includes three observation well profiles (transects) orientated along streamlines with typical redox gradients. Due to the operation of the water works and the damming of the storage lake, the aquifer of the *Insel Hengsen* is divided in three areas with significantly different redox conditions:

1. Some areas are influenced by artificial recharge of pre-filtered, oxic surface water. Here, oxidising redox conditions characterise the hydrochemical processes in the aquifer. Transect III is placed in this area (Figure 13.1).

2. Areas distant from artificial recharge are exclusively influenced by bank filtration. Here, a distinct change from oxidising to strictly reducing redox conditions can be observed. Transect I is placed in one of these areas (Figure 13.1).
3. In areas where artificial recharge and bank filtration are in close contact, the formation of a transient zone can be observed. Here, depending on the operation of the water works, oxidising and reducing hydrochemical conditions frequently change. Most of the times, these changes are faster than the development of chemical equilibria between the liquid and the solid phases of the aquifer. Transect II is placed in one of these areas (Figure 13.1).

13.3 Observed Redox Conditions

In the course of the investigations, between 100 and 160 E_H-values were measured in each observation well of the transects. In Figure 13.2 the development of the groundwater redox conditions during the infiltration and underground passage in the three transects is shown. Presented are the mean values per observation well and the range band of ±1 standard deviation.

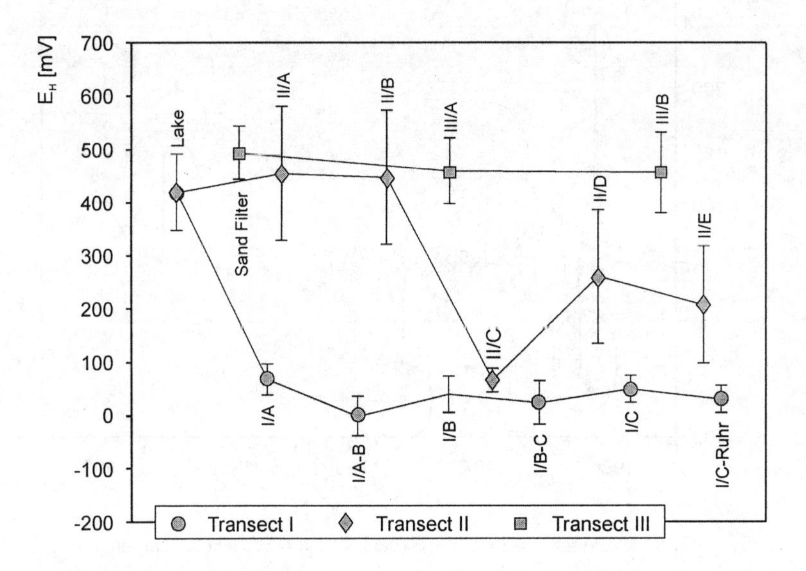

Fig. 13.2: Redox conditions along Transects I (bank filtrate), II (bank filtrate and artificial recharge) and III (artificial recharge). Data points represent mean values and standard deviations. See Figure 13.1 for well locations.

Figure 13.2 clearly shows the transition of the hydrochemical environment from oxidising to reducing conditions during bank filtration in Transect I. The surface water with E_H-values of about 400 mV changes its E_H-values to around 0 mV due to the reduction of oxygen, nitrate, iron, manganese and sulfate during infiltration and underground passage. In the course of Transect I the E_H-value stays relatively constant with only minor changes. The range band of the mean values ±1 standard deviation is relatively small in Transect I which is not influenced by the operation of the water works and shows only very slight changes in the hydraulic boundary conditions. Seasonal changes in the surface water composition mainly determine the changes in E_H-values.

Figure 13.2 also illustrates the typical development of redox potentials during artificial recharge by slow-sand filters in Transect III. The pre-filtered, oxygen enriched surface water with a E_H-value of about 500 mV is only slightly reduced to E_H-values of about 450 mV, mainly due to the reduction of oxygen. The range band of the mean values is broader than that of Transect I. The intermittent operation of the slow-sand filter and seasonal changes in the composition of the surface water have the main influence on changes in E_H-values.

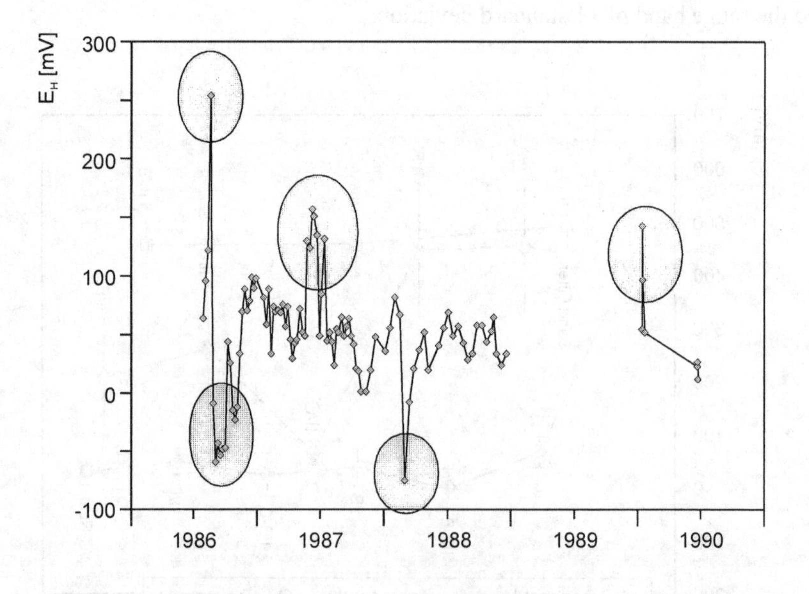

Fig. 13.3: Redox conditions in well I/C (bank filtrate). See text for explanations.

Transect II is characterised by a combination of bank filtration and artificial recharge. The transition of oxidising hydrochemical conditions with E_H-values of about 400 mV toward reducing conditions during bank filtration up to observation well II/C with values of about 100 mV is obvious. From that point, the episodic influence of less reduced artificially recharged water is clearly visible. The mean

E_H-values resume to about 250 mV. Transect II shows the most pronounced range band of the mean values plus and minus one standard deviation. The changes in E_H-values caused by seasonal changes in the surface water composition are strongly confirmed by the occasional influence of less reduced, artificially recharged groundwater.

13.4 Interpretation Problems

As far as only mean values or a statistical range band of E_H-values are regarded, the general appearance of the development of the observed redox conditions in the different transects seems logical. However, a more detailed look at single values and their relation to the series of measured values shows some discrepancies. For example the single E_H-values of observation wells I/C (strongly reducing conditions), III/A (strongly oxidising conditions) and II/E (influence of bank filtrate and artificial recharge) are shown in Figures 13.3-5. In these figures, values that considerably deviate from the general range band are marked with a circle. These values are neither explainable with seasonal changes in the composition of the surface water nor with the operation of the water works.

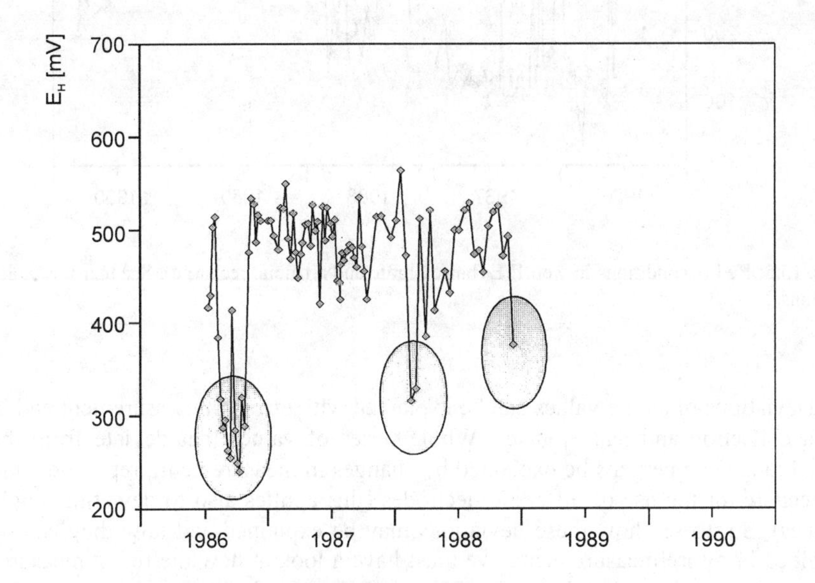

Fig. 13.4: Redox conditions in well III/B (artificial recharge). See text for explanations.

Two series of E_H-values in observation well I/C (Figure 13.3) from autumn 1986 and summer 1987 deviate by about 100 mV from the usual range band.

Some single values in 1986, 1988 and 1990 show even larger deviations from the usual range band. In autumn 1986, a similar deviation of a series of values was observed also in observation well III/B (Figure 13.4). Here, the E_H-values were about 200 mV lower than usual. In the same well deviations of single values can also be recognised. The tracing of doubtful E_H-values in observation well II/E (Figure 13.5), situated in an area of Transect II which is influenced by bank filtrate and artificial recharge, is much more difficult. The changing influence of anaerobic bank filtrate and aerobic recharge by itself produces a broad range of E_H-values. So even large deviations of single values can be explained with this process.

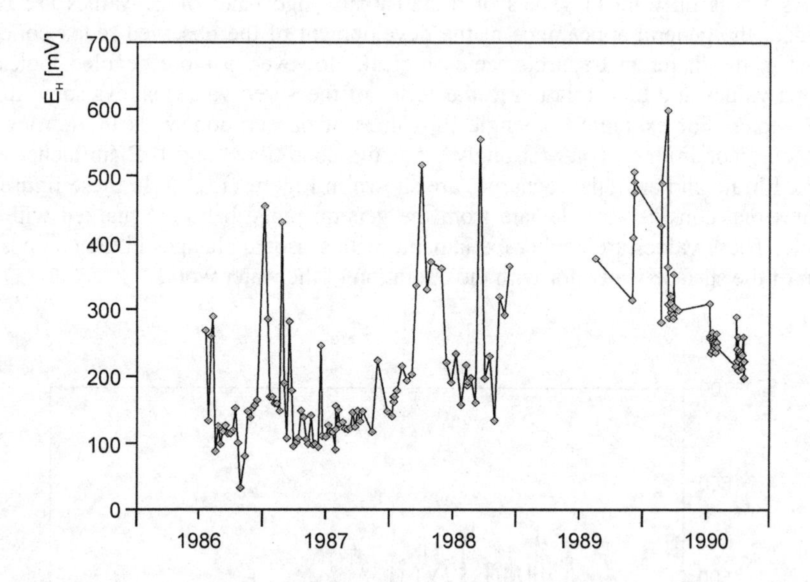

Fig. 13.5: Redox conditions in well II/E (bank filtrate and artificial recharge). See text for explanations.

Deviations of single values can be explained with errors in measurement and in data collection and data transfer. Whole series of values that deviate from the usual range can perhaps be explained by changes in measurement preparation and procedure, or the use of different electrodes (this applies also to deviating single values). So, to see how these deviations may be explained and how they can be avoided in future measurements, we must have a look at how the redox measurements were carried out and what errors/deviations are to be expected.

13.5 Groundwater Sampling and Redox Measurement

It is well known that observation wells yield a water chemistry different from production wells. The main reason for this effect is the presence of stagnant water above the well screen. Therefore, observation wells have to be purged prior to groundwater sampling. Estimates for the number of well volumes to be emptied vary between 2 and 10 (BARBER & DAVIS, 1987; ROBIN & GILLHAM, 1987; STUYFZAND, 1983). Sampling for special parameters like colloidal and suspended particles can require an extended purging up to 700 well volumes (HOFMANN & SCHÖTTLER, 1998).

At the test site *Insel Hengsen*, groundwater samples were taken with a submersible pump at a varying flow rate between 1 and 15 l/min. Electrode measurements for E_H, pH, O_2 and electric conductivity (EC) were carried out in an in-line flow cell. The bypass flow rate for the electrode measurements was adjusted to 0.2 – 1 l/min. In general, observation wells were purged for 30 minutes (three well volumes). Extended purging of up to 5 days was only necessary for the sampling of colloidal and suspended particles.

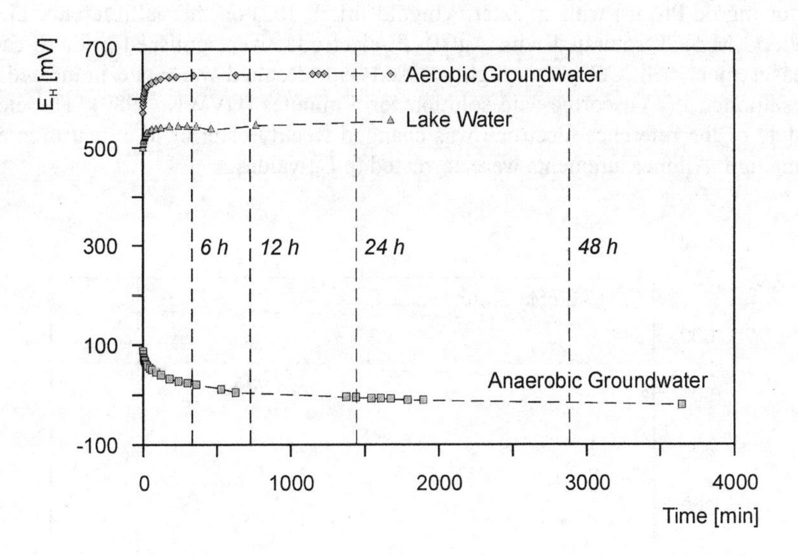

Fig. 13.6: Temporal development of redox conditions in various water types.

Theoretically, the E_H determines the distribution of all redox equilibria in a similar way as the pH expresses the distribution of all acid-base equilibria. However, the E_H cannot be measured unambiguously in most natural waters. E_H-measurements are made with an inert-Pt-electrode against a standard electrode of a known potential. Therefore all measurements should be corrected to the potential relative to the standard hydrogen electrode.

Although waters from aerobic environments yield higher E_H-values than waters from anaerobic environments, it proved to be very difficult to obtain a meaningful interpretation in sense of the NERNST-equation (APPELO & POSTMA, 1994). Discrepancies between the calculated and measured data sets are very large. Another important problem is the fact that many redox couples are not electroactive. For example, no reversible electrode potential can be established for NO_3^--NO_2^--NH_4^+-H_2S or CH_4-CO_2 (STUMM & MORGAN, 1981). Beside the lack of equilibrium between different redox couples in a water sample, analytical difficulties are of major importance. They include the lack of electroactivity at the Pt-surface, mixing potentials, and poisoning of the electrode. An example is precipitation of Fe-oxihydroxides on the Pt-surface in anaerobic environments because of O_2 adsorbed on the electrode surface (DOYLE, 1968). It will lead to a reduction in the electron-exchange rate or in the effective area of the electrode, and thus alter the exchange current of the electrode.

In order to minimise theses effects and to improve the reproducibility of the measurements, an identical pre-treatment of redox-electrodes prior to each measurement is highly recommended. In the following experiments and during sampling campaigns at *Insel Hengsen* test site, we used Pt-electrodes (Ingold Pt4805-S7 or Ingold Pt-pin) with an extern (Ingold InLab 302) or internal reference electrode (3 M KCl, saturated with AgCl). Pt-electrodes were polished prior to each measurement with CeO_2, thoroughly rinsed with desalted water and immersed in an saturated L(+) ascorbic acid solution for 5 minutes (DVWK, 1989). The electrolyte of the reference electrode was changed weekly or prior to a measurement campaign. All measurements were corrected to E_H-values.

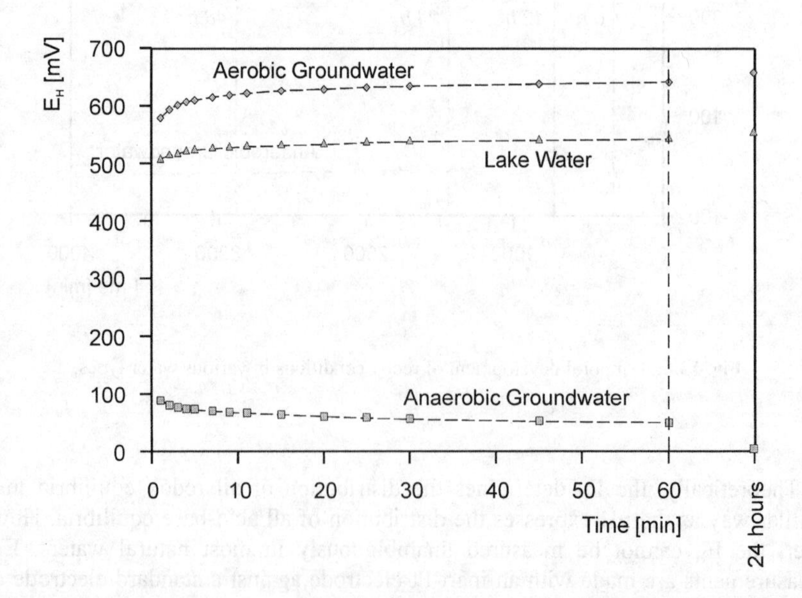

Fig. 13.7: Temporal development of redox conditions in various water types during the first 60 minutes.

13.6 Kinetics of Redox Measurements

As stressed earlier, groundwater sampling in observation wells requires the exchange of the stagnant water above the well screen. In general, groundwater samples were taken after 30 minutes of purging at the test site *Insel Hengsen.* After this time O_2, pH, EC and temperature were constant. The redox potential showed only a very slight drift towards smaller E_H-values. During sampling campaigns for special parameters (colloidal and suspended particles) significant differences between 30 minutes, 1 hour and 24 hour readings were noticed. In order to clarify these observations, several experiments with anaerobic and aerobic groundwater, lake water and three different redox electrodes were performed over a measurement period up to 60 hours.

In Figure 13.6, redox measurements in three different water types, i.e. anaerobic groundwater, aerobic groundwater and water from an aerobic lake, are plotted against time. Measurements were performed with a Pt-electrode with an internal reference electrode. The initial E_H-values correspond to the different environments. E_H-values are +580 mV for aerobic groundwater (*AG*), +500 mV for the aerobic lake water (*AL*) and +90 mV for the anaerobic groundwater (*AnG*).

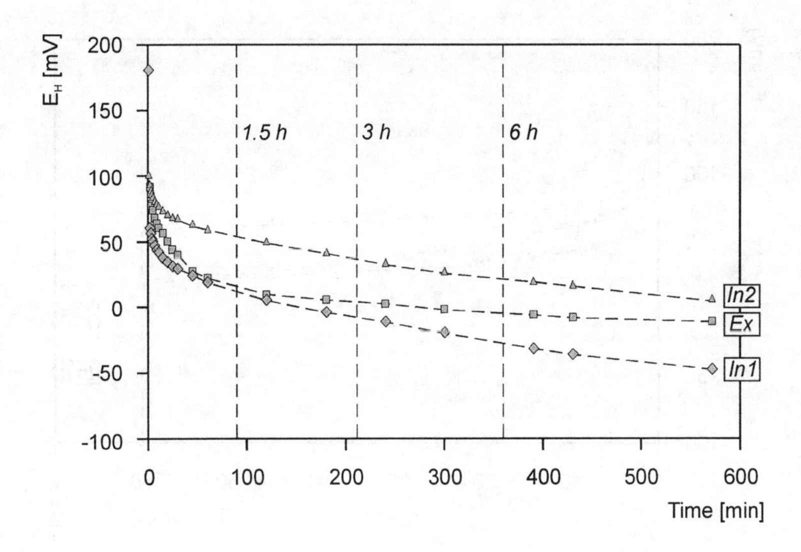

Fig. 13.8: Temporal development of redox conditions in anaerobic groundwater (*AnG*) measured with three different electrodes.

A significant drift is obvious in the first hour, but does not even decline to zero after more than 24 h, where E_H-values reach +650 mV (*AG*), +560 mV (*AL*) and -5 mV (*AnG*). In Figure 13.7, the first 60 minutes of the data from Figure 13.6 are displayed. It is evident that values at 30 minutes do not reflect the final conditions, especially at anaerobic environments. The disparity between 30 minutes reading and 24 h reading is 30 mV (*AG* and *AL*) and 65 mV for *AnG*. Due to the large discrepancies, anaerobic groundwater measurements were continued to up to 60 hours (Figure 13.6). The E_H-value dropped continuously to –50 mV (*AnG*) and a constant value could not be reached. The disagreement between 30 minutes reading and 60 h reading was 110 mV.

Figure 13.8 displays redox measurements in the same environment (*AnG*), but which were carried out with three different electrodes. Two Pt-electrodes with an internal reference electrode (*In1* and *In2*) were used together with a Pt-electrode with an external reference electrode (*Ex*). The three electrodes display an exponential drop of the E_H-values with a steep part during the first hour. Final E_H-values are +5 mV (*In2*), -15 mV (*Ex*) and –50 mV (*In1*). The Pt-electrode with an external reference electrode shows a steep decrease during the first hour reaching the same values as *In1* during the first hour. Subsequently, the decrease is slower and values are closer to the values of *In2* after 9 h. This elucidates that the decrease of each electrode is different and can not be predicted. The E_H-values after 30 minutes in Figure 13.9 differ from those after 9 h about 60 mV (*In2*), 50 mV (*Ex*) and 80 mV (*In1*).

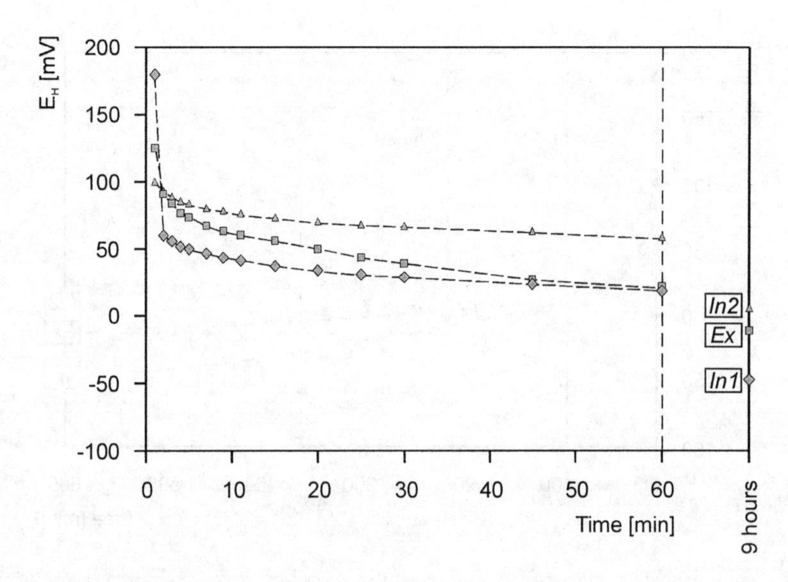

Fig. 13.9: Temporal development of redox conditions in anaerobic groundwater (*AnG*) measured with three different electrodes during the first 60 minutes.

In Figure 13.10 the same electrodes are tested in aerobic groundwater. While the disagreement between *In1* and *In2* was as much as 55 mV in anaerobic groundwater, differences in the aerobic environment are very small (5 mV). Also,

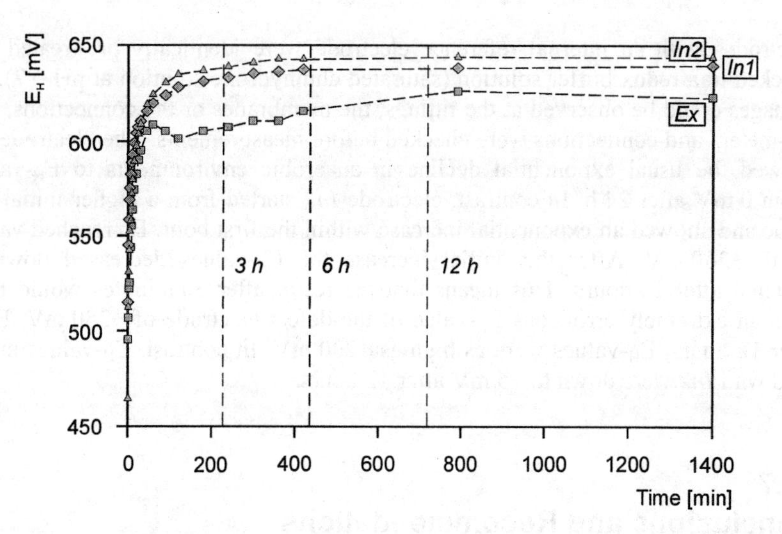

Fig. 13.10: Temporal development of redox conditions in aerobic groundwater (*AG*) measured with three different electrodes.

the exponential increase of the two electrodes nearly fits over the 24 h measurement period. In contrast, the values from *Ex* vary significantly from *In1* and *In2* between 1 h and 6 h, reaching a value 20 mV below the values of *In1* and *In2*. The variance between 30 minute values and 24 h values was 35 mV for *In1* and *In2* electrodes and 25 mV for the *Ex* electrode.

Apart from the disagreements between different electrodes observed in the previous experiments, extreme effects can be observed in Figure 13.11. Two Pt-

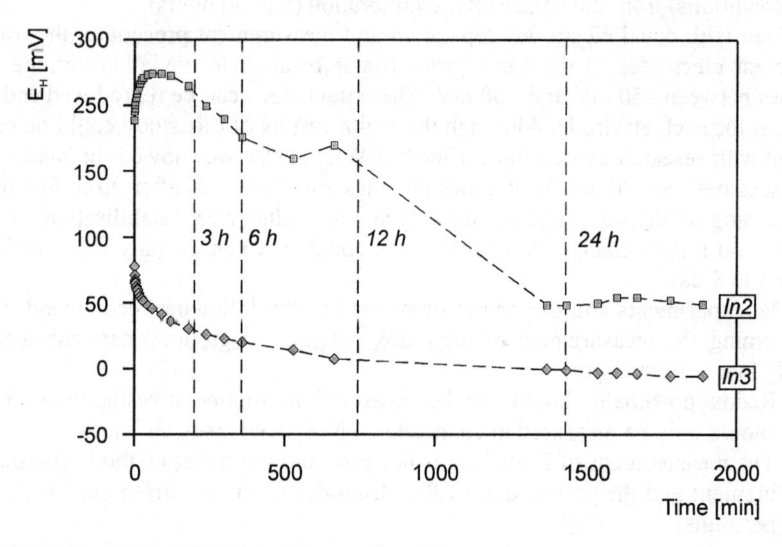

Fig. 13.11: Erroneous redox measurement in anaerobic groundwater (*AnG*) due to the defect electrode *In3*.

electrodes with an internal reference electrode were identically pre-treated and checked in a redox buffer solution (saturated chinhydrone solution at pH = 7). No damages could be observed at the fittings, the membranes or the connections. The pH-meters and connections were checked before measurements. The electrode *In2* showed the usual exponential decline in anaerobic environments to E_H-values about 0 mV after 24 h. In contrast, electrode *In3* started from a higher initial E_H-value and showed an exponential increase within the first hour. E_H reached values up to +260 mV. After this initial increase the E_H-values decreased down to +50 mV after 24 hours. This means that the result after 30 minutes would have been an extremely erroneous E_H-value of the defect electrode of +250 mV. Even after 12 hours, E_H-values were as high as +200 mV. In contrast, E_H-values measured with *In2* were down to +5 mV after 12 hours.

13.7 Conclusions and Recommendations

The experiments carried out to quantify variations between different redox electrodes and the kinetics of electrode adjustment have shown a broad range of possible results and a considerable time dependency. Even under laboratory conditions with identical electrode preparation and measurement procedure, the use of different electrodes and varying measurement times resulted in a tremendously broad range of E_H-values. Under real field conditions, these ranges must be even larger.

The electrodes tested in natural groundwater showed a slow kinetic of electrode adjustment up to 60 hours. Due to the resulting lack of electroactivity at the Pt-surface and due to poisoning of the electrode this is typical for groundwater with a relatively low ionic strength. Values obtained after only 20 to 30 minutes measurement time differed between +50 mV (aerobic conditions) to –100 mV (anaerobic conditions) from the values after equilibration (9 to 60 hours).

Even with equal electrode preparation and measurement procedure, the use of different electrodes in the same groundwater resulted in deviations of the E_H-values between –50 mV and +50 mV. These deviations can be reproduced and are typical for each electrode. Although the major results of this study could be compared with research carried out earlier (DVWK, 1989), we showed the major differences between 20 and 30 minutes readings and E_H-values after up to 60 hours. Depending on the water and electrode type, the reading after "stabilisation" of the E_H, i.e. no further change than 1 mV/5min, could vary significantly from readings after 1 to 3 days.

The experiments and the observations lead to the following recommendations concerning the measurement of the redox potential in groundwater with a relatively low ionic strength:

- Redox potentials should not be measured in routine investigations. They should only be measured in connection with special research issues.
- The measurement of E_H-values is not easy and not a fast method. The measurement and the preparation of the electrodes must be carried out by trained personnel.

- Measurements should always be carried out with the same electrode. Data sets measured with different electrodes are difficult to compare.
- An appropriate time for the adjustment of the electrode must be allowed. At the start of a measurement campaign tests should be carried out to determine the kinetics of electrode adjustment in the particular environment (several hours).
- A plausibility test with other parameters is always necessary (e.g. oxygen, or calculated E_H-values on the basis of analysed parameters) to recognise serious malfunctions of the electrode.
- Redox buffers (chinhydrone) do not always help to identify malfunctions of an electrode. They just prove that contact between electrode and the measurement device is given.

13.8 Acknowledgements

The results shown where obtained in research projects funded by Deutsche Forschungsgemeinschaft (DFG), the European Union (EU, contract ENV-CT95-0071) and Deutscher Verein des Gas- und Wasserfachs (DVGW).

13.9 References

APPELO, C.A.J. & POTSMA, D. (1994): Geochemistry, groundwater and pollution. Balkema, Rotterdam.

BARBER, C. & DAVIS, G.B. (1987): Representative sampling of ground water from short-screened boreholes. GroundWater 25: 581-587.

DOYLE, R.W. (1968): The origin of the ferrous ion-ferric oxide Nernst potential in environments containing ferrous iron. Am. J. Sc. 266: 840-859.

DVWK (Deutscher Verband für Wasserwirtschaft und Kulturbau e.V.) (1989): Grundwasser: Redoxpotential und Probenahmegeräte. DVWK-Schriften, Heft 84, Verlag Paul Parey, Hamburg.

HENCKE, J. & SCHULZ, H.D. (1998): Laborsäulenversuche zum Redoxverhalten anorganischer Spurenstoffe. Z. dt. geol. Ges., 148/3-4: 369-387.

HOFMANN, T. & SCHÖTTLER, U. (1998): Microparticle facilitated transport of contaminants during artificial groundwater recharge. In: PETERS et al. (eds.): Artificial recharge of Groundwater, Balkema, Rotterdam, 205-210.

HOFMANN, T. & SCHULTE-EBBERT, U. (1997): Mikropartikel und künstliche Grundwasseranreicherung, 2. Deutsch-Niederländischer Workshop Künstliche Grundwasseranreicherung, DVGW-Schriftenreihe Wasser 90: 115-130.

ROBIN, M.J.L. & GILLHAM, R.W. (1987): Field evaluation of well purging procedures. Groundwater Monitoring Review 7: 85-93.

SCHÖTTLER, U. & SCHULTE-EBBERT, U. (Eds.) (1995): Schadstoffe im Grundwasser, Bd. 3 Verhalten von Schadstoffen bei der Infiltration von Oberflächenwasser am Beispiel des Untersuchungsgebiets *Insel Hengsen* im Ruhrtal bei Schwerte. Deutsche Forschungsgemeinschaft (DFG), Bonn.

SCHULTE-EBBERT, U. & SCHÖTTLER, U. (1998): Transport von polycyclischen aromatischen Kohlenwasserstoffen (PAK) bei der Untergrundpassage. Interims report research project Scho 337/6. Deutsche Forschungsgemeinschaft (DFG), Bonn.

STUMM, W. & MORGAN, J.J. (1981): Aquatic Chemistry. 2nd ed., Wiley-Interscience, New York.

STUYFZAND, P.J. (1983): Important sources of errors in sampling groundwater from multilevel samplers (in Dutch). H_2O 16: 87-95.

Chapter 14

Redox Reactions, Multi-Component Stability Diagrams and Isotopic Investigations in Sulfur- and Iron-Dominated Groundwater Systems

F. Wisotzky

14.1 Introduction

Redox reactions of sulfur are very common in groundwater, surface water and in the marine environment (NORDSTROM et al., 1979; FROELICH et al., 1979; LEUCHS, 1988; ALPERS & BLOWES, 1994; VAN BERK & WISOTZKY, 1995; STUMM & MORGAN, 1996). The input of an oxidant (O_2, NO_3^-, Fe(III)) into a system which contains sulfide minerals leads to mobilising reactions (increase of mineralisation and decrease of pH). Thus, the reaction of oxygen or nitrate with sulfide-containing soils is responsible for sulfuric acid release and increased mobility of metals. Pyrite (FeS_2) as an important sulfide mineral can be found in metallic ores, black shales and in overburden sediments of lignite and hard coal deposits. Pyrite oxidation processes include redox reactions of sulfur and iron. These reactions are of high importance for the groundwater quality in lignite mining areas where there are massive occurrences of groundwater in porous media. High sulfate and iron concentrations in the groundwater of the dump aquifers corroborate the dominance of pyrite oxidation in this system. This characterisation is necessary to check the plausibility of redox measurements in the investigated system.

The input of organic pollution can also be degraded by ferric iron and sulfate reduction, if the organic substances are degradable by sulfate- and/or ferric iron-reducing bacteria. Ferrous iron is involved in the formation of iron sulfide minerals after ferric iron and sulfate reduction. The reduction of Fe(III)-hydroxide minerals proceeds at a higher redox potential than the sulfate reduction (FROELICH et al., 1979; LEUCHS, 1988; STUMM & MORGAN, 1996) and has to be taken into account. As an example for sulfate reduction in aquifers, a BTEX pollution (benzene, toluene, ethylbenzene and xylenes) at a former gasworks is presented (see also WISOTZKY & ECKERT, 1997; SCHMITT et al., 1996). The dominance of sulfate reduction in the investigated aquifer is proven by water chemistry (sulfate, BTEX), the analysis of solid composition of the sediments (sulfide content, TIC-content) and isotopic data ($\delta^{18}O\text{-}SO_4^{2-}$, $\delta^{34}S\text{-}SO_4^{2-}$). A plausibility test of redox potential measurements is made possible by this proof.

14.2 Methods

14.2.1 Locations of the Study Area

The investigation was performed at two different sites. Water samples from the dump aquifer Berrenrath of the Rhineland Lignite Mining Area predominately contain pyrite oxidation products. The area of the overburden disposal site covers approximately 100 ha and measures an average thickness of 40 m (20 m unsaturated). The dump aquifer contains unlithified sediments displaying a variable sediment composition, especially with respect to various pyritic-sulfur (FeS_2-S).

The second investigation area of a BTEX-polluted groundwater is located in the city of Düsseldorf (Germany). The aquifer consists of late Pleistocene fluvial deposits with medium to coarse grained sand and gravel from the river Rhine. At the former gasworks plant massive soil and groundwater contamination was detected in the shallow unconfined aquifer.

14.2.2 Sediment and Water Analyses

Dissolved calcium and sulfate concentrations were determined by ion chromatography (DIONEX DX 500). Iron concentrations were measured by ICP-AES (PHILLIPS PU 7000). Redox potential and pH were detected on-site in a flow-through cell with the WTW (pH 196). For redox measurements a platinum electrode was used with an Ag/AgCl-reference electrode (Pt 4805) in a 3 mol/L KCl-solution (METTLER). The measured values were corrected to the standard hydrogen electrode (SHE). The sediments of 11 core drillings (position see Figure 14.2) of the BTEX-polluted aquifer were analysed as to the total inorganic carbon content (TIC) and sulfide content. The core samples were brought into nitrogen-gas flooded plastic bags and were frozen at -20°C until preparation. After freeze drying the samples were ground and homogenised. The TIC which represents the carbonate-carbon content was coulometrically measured after addition of heated perchloric acid (90°C) as carbon dioxide (STRÖHLEIN, Coulomat 702). The sulfide content was also measured coulometrically (Coulomat 702) after heating up to 550 °C following BRUMSACK (1981), LEUCHS (1988) and WISOTZKY (1994).

14.2.3 Analysis of Isotopes

Analysis of isotopes of $\delta^{18}O\text{-}SO_4^{2-}$ and $\delta^{34}S\text{-}SO_4^{2-}$ were performed in the BTEX-polluted aquifer. The measurements were carried out with the mass spectrometers MAT 251 and Delta S (FINNIGAN) at the Institute of Sediment and Isotope Geology (Ruhr-University Bochum, Germany). The isotopic values are given in del notation as deviation of isotope ratios from the standards SMOW and CDT (SMOW: $\delta^{18}O\text{-}SO_4^{2-}$; CDT: $\delta^{34}S\text{-}SO_4^{2-}$). Lighter isotopes ($^{32}S$ and ^{16}O) are preferentially used in the process of bacterial sulfate reduction (CHAMBERS & TRUDINGER, 1979). This leads to an enrichment of lighter isotopes in reduction products (e.g. H_2S, HS^-, FeS(s), FeS_2(s)) until the depletion of sulfate is completed. An enrichment of heavier isotopes (^{34}S and ^{18}O) in the remaining sulfate of the groundwater is the result.

14.2.4 Multi-Component Stability Diagrams (pH-pε-Diagrams)

Typical groundwater analyses were used to calculate stability diagrams (pH-pε-diagrams) of iron minerals by means of the newly developed computer code (Dept. Applied Geology, Ruhr-University Bochum, Germany) based on PHREEQM (APPELO & WILLEMSEN, 1987). The calculations were performed with data from water analyses of a lignite mining dump and from groundwater analyses where samples were collected from a BTEX pollution at a former gasworks. Fe(III)-hydroxide, siderite ($FeCO_3$(s)) and pyrite (FeS_2(s)) were used as the relevant iron minerals in the calculations. The solubility product of $Fe(OH)_3$(s) from the thermodynamic data set of the computer code PHREEQM represents the Fe(III)-hydroxide mineral phase ferrihydrite (see CORNELL & SCHWERTMANN, 1996). The following solid phase constants were used:

$$Fe(OH)_3 \leftrightarrow Fe^{3+} + 3H_2O - 3H^+ \qquad \log k = 4.891$$
$$FeS_2 \leftrightarrow Fe^{2+} + 2HS^- - 2H^+ - 2e^- \qquad \log k = -18.48$$
$$FeCO_3 \leftrightarrow Fe^{2+} + CO_3^{2-} \qquad \log k = -10.89$$

Goethite and hematite were excluded from the study because ferrihydrite was used as the relevant ferric iron hydroxide mineral. The new developed stability diagrams were calculated by variation of pH and pε, starting from measured values and using all complexation reactions of the computer code PHREEQM, corrected for ionic strength effects and for real groundwater temperatures. In comparison to conventional stability diagrams with only a few components (GARRELS & CHRIST, 1965; BROOKINS, 1987; STUMM & MORGAN, 1996), the diagrams calculated using the newly computer code based on PHREEQM, are multi-component stability diagrams for realistic water samples. Maximally five components are only used for constructing conventional stability diagrams. The pH-pε-diagrams presented here were calculated with the measured concentrations of all inorganic components in groundwater (Ca^{2+}, Mg^{2+}, Na^{+}, K^{+}, $Fe^{2+/3+}$, Mn^{2+}, Al^{3+}, Cl^{-}, SO_4^{2-}, NO_3^{-}, HCO_3^{-}, H_2O). The total concentrations remained constant in the calculations.

14.3 Results

14.3.1 Mobilising Reactions of Sulfur and Iron

Iron sulfide oxidation can be subdivided into the oxidation of sulfide-sulfur (first oxidation step) and the oxidation of ferrous-iron (second oxidation step; Table 14.1).

Tab. 14.1: The different steps of pyrite oxidation including partial reactions as mobilising reactions (increase of mineralisation and possible acidification).

First oxidation step of pyrite oxidation (sulfide oxidation)		
$FeS_2(s) + 14Fe^{3+} + 8H_2O$	→	$15Fe^{2+} + 2SO_4^{2-} + 16H^{+}$
$14Fe^{2+} + 7/2O_2 + 14H^{+}$	→	$14Fe^{3+} + 7H_2O$
$FeS_2(s) + 7/2O_2 + H_2O$	→	$Fe^{2+} + 2SO_4^{2-} + 2H^{+}$
Second oxidation step of pyrite oxidation (ferrous-iron oxidation)		
$Fe^{2+} + 1/4O_2 + H^{+}$	→	$Fe^{3+} + 1/2H_2O$
$Fe^{3+} + 3H_2O$	↔	$Fe(OH)_3 + 3H^{+}$
$Fe^{2+} + 1/4O_2 + 2.5H_2O$	→	$Fe(OH)_3 + 2H^{+}$
Complete pyrite oxidation (sulfide and ferrous-iron oxidation)		
$FeS_2(s) + 15/4O_2 + 3.5H_2O$	→	$Fe(OH)_3 + 2SO_4^{2-} + 4H^{+}$

The sulfide oxidation and the resulting "sulfide acidity" mainly react in the groundwater of the dumps. The second oxidation step (ferrous-iron oxidation) will be responsible for further acidification, if the water leaves the aquifer (at natural groundwater discharges or residual lakes) and comes into contact with oxygen (Table 14.2). The complete oxidation of sulfide-sulfur and ferrous-iron of pyrite leads to a release of 4 mol protons per mol pyrite (Table 14.1). The two oxidation steps of pyrite oxidation in dump aquifer and residual lake systems proceed in a time-dependent manner and locally separated, which is due to the lack of oxygen. This oxygen shortage occurs in open pit mines ("primary pyrite oxidation") and in recultivated dumps ("secondary pyrite oxidation"; see VAN BERK & WISOTZKY, 1995; WISOTZKY, 1998a). Therefore, only 14% of the pyritic sulfur content in the whole overburden material of the lignite mine Garzweiler is oxidised, and equivalent quantities of sulfate, ferrous-iron and protons are produced (WISOTZKY, 1994). Besides the oxidation kinetics the slow oxygen transport in the soil gas by air convection and diffusion (see VAN BERK & WISOTZKY, 1995) is responsible for the low content of oxidised pyritic sulfur in dump aquifers. Chemical groundwater analysis at the overburden disposal site are revealed that only the first oxidation step (sulfide oxidation) is realised, which is due to oxygen deficiency. Therefore, the pH of the groundwater in dumps is higher (pH >3.5) than the pH of the surface water in residual lakes that essentially consist of former groundwater (mostly pH = 2 to 4; see PIETSCH, 1979; FRIESE et al., 1998). Pyrite is completely oxidised only in certain sections of the dumps attributable to a low iron sulfide content or a longer time of exposure to the oxidising atmosphere. There, a complete oxidation of sulfide and ferrous-iron does occur.

Tab. 14.2: Changes of the water chemistry of the anoxic groundwater and after aeration (O_2-contact) of natural groundwater discharges of the three different parts of the lignite dump aquifer Berrenrath.

	pH	*pε*	Fe^{tot} [mM]	SO_4^{2-} [mM]	HCO_3^- [mM]	Ca^{2+} [mM]
Low pyrite content in sediments (<0.15 wt% FeS_2-S)						
Groundwater	6.05	2.7	3.85	8.90	12.75	6.17
O_2-contact	6.33	9.3	0.82	9.17	0.84	6.17
Intermediate pyrite content in sediments (≈0.15 wt% FeS_2-S)						
Groundwater	5.88	4.2	6.81	22.92	2.30	12.35
O_2-contact	4.00	10.1	3.58	21.88	<0.02	12.59
High pyrite content in sediments (≈0.23 wt% FeS_2-S)						
Groundwater	4.19	6.4	57.53	62.53	<0.02	10.32
O_2-contact	2.45	10.6	14.33	31.30	<0.02	10.32

In contrast to surface waters, the groundwater of the dump mainly contains dissolved iron in its reduced form (Fe(II)). Further acidification (decrease of pH) of the water in contact with atmospheric oxygen indirectly indicates the lack of oxygen in dump aquifers. Without oxygen shortage there has to be a comparable low pH in groundwater and in oxidised water of natural groundwater discharges or residual lakes. The decrease of pH after aeration is mainly determined by the Fe(II)-concentration and the pH-buffer capacity of the groundwater. Table 14.2 shows the change in water chemistry before (groundwater) and after O_2-contact (natural groundwater discharges) of three different parts (different contents of pyritic sulfur) belonging to the dump aquifer Berrenrath. Measurements listed in Table 14.2 show that the redox potential increased by at least 4 pε units (>200 mV) caused by O_2-contact at natural groundwater discharges. The resulting precipitation of the Fe(III)-hydroxide mineral diminishes the dissolved iron concentration and mostly leads to a decrease of pH (at ≥0.15 wt% FeS_2-S in Table 14.2). One part of the dump aquifer Berrenrath revealing a low pyrite content (at <0.15 wt% FeS_2-S in Table 14.2) is exceptional. There, carbon dioxide degassing leads to a pH increase despite the oxidation and precipitation of the Fe(III)-hydroxide mineral (Table 14.2). A high Fe(II) concentration in the groundwater causes an intensive acidification (intermediate and high pyrite content in Table 14.2). The calcium and sulfate concentrations mostly remain constant. Aeration decrease the sulfate concentrations only in the part of the overburden dump which contained a concentration of high pyrite. This finding is perhaps related to the formation of a Fe(III)-hydroxysulfate mineral like Jarosite or Schwertmanite (see CORNELL & SCHWERTMANN, 1996; BIGHAM et al., 1996).

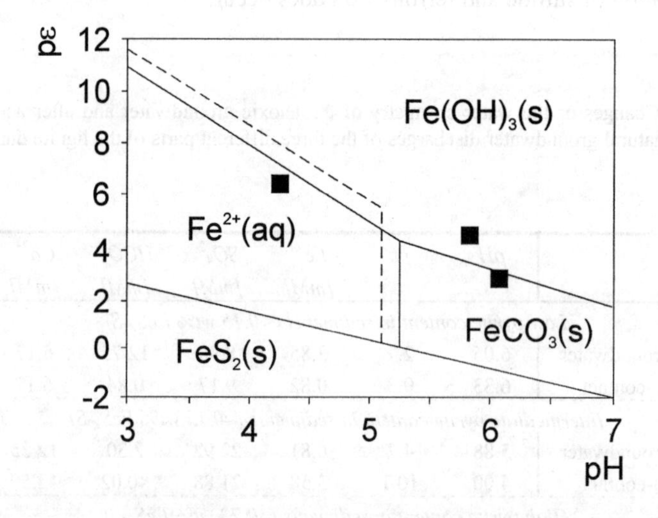

Fig. 14.1: Stability diagrams for different iron-bearing phases and species of three groundwater analyses from the dump aquifer Berrenrath calculated with the computer code based on PHREEQM (solid lines represent water analysis with pH 4.19; dashed lines represent two other analyses at pH = 6.05 or 5.88 (see Table 14.2; $Fe(OH)_3$ represents a Fe(III)-hydroxide mineral).

The measured redox potential is sensitive to the absence (groundwater) or presence (natural groundwater discharge) of oxygen, or to any change of the redox state of the dissolved iron (Fe(II)/Fe(III)). NORDSTROM et al. (1979) have found an equilibrium of the redox couple Fe^{2+}/Fe^{3+} and/or $Fe^{2+}/Fe(OH)_3(s)$ in a strong acid environment (pH = 1-2) with pε >4. Despite the input of oxygen, a shift to very high positive values of the redox couple O_2/H_2O was not measured. This can be explained by the fact that the exchange current of O_2 at the electrode is too low (SIGG & STUMM, 1994), or that it results from the remaining Fe^{2+} concentration in water (Table 14.2). For testing the plausibility of redox measurements, three groundwater analyses (Table 14.2) were used to calculate stability diagrams (pH-pε-diagrams; Figure 14.1) with the new computer code based on PHREEQM. The solid lines in Figure 14.1 represent the water analysis with a pH of 4.19 (Table 14.2), whereas the dashed lines represent the compositions of two other analyses shown in Table 14.2. The resulting pH-pε-diagrams of these three groundwater analyses are similar.

The pH-pε-values from the groundwater analyses of the dump aquifer Berrenrath (Table 14.2) are located on the stability boundary between different Fe(II)- and Fe(III)-phases or species (Figure 14.1). The measured redox potentials seem to be determined by the Fe^{2+}/Fe^{3+} or the $Fe^{2+}/Fe(OH)_3(s)$ redox couple.

The stability field of pyrite occurs at lower redox values than those measured in the groundwater. Metastable FeS would be formed first within the stability field of pyrite. Therefore, the formation of iron sulfide minerals by means of intensive sulfate reduction in the investigated dump aquifer is unlikely (see WISOTZKY, 1998c). But the remaining primary pyrite of the sediments (not oxidised during mining activity) also does not seem to have an effect on the measured redox values in groundwater. Thus, a complete redox equilibrium is not reached in the dump aquifer. The changed pH-pε-values in the three natural groundwater discharges (Table 14.2) are mostly located at the Fe(II)/Fe(III) stability boundary (not shown). Only the analysis demonstrating the lowest iron concentration in groundwater (Table 14.2) completely moves into the Fe(III)-hydroxide stability field (not shown) due to oxygen contact.

The calculations (Figure 14.1) applying the measured redox potentials of groundwater point out that the pε-value is a sensitive and useful parameter in groundwater systems where sulfur and iron constitute the main redox elements. However, only a partial equilibrium (see STUMM & MORGAN, 1996) is reached in the groundwater from dump aquifers. Thus, the remaining pyrite content in the sediments is not in equilibrium with the measured redox potential values.

14.3.2
Immobilising Reactions of Sulfur and Iron

Besides mobilising reactions of sulfur and iron (see Chapter 14.3.1), immobilising sulfate and Fe(III)-reduction processes are relevant in aquatic systems (e.g. FROELICH et al., 1979; LEUCHS, 1988; SIGG & STUMM, 1994; STUMM & MORGAN, 1996; WISOTZKY & ECKERT, 1997).

Here, sulfate reduction is coupled with the oxidative degradation of benzene (Table 14.3). In general, groundwater pollution problems with mono-aromatic hydrocarbons are observed quite often (HOFFMANN, 1993; BARBARO et al., 1992; BARKER et al., 1987, MAJOR et al., 1988; PATTERSON et al., 1993; WISOTZKY, 1998b) which is due to a comparatively high solubility of the mono-aromatic

Tab. 14.3: Possible benzene degradation reactions as an example for BTEX (benzene, toluene, ethylbenzene and xylenes) degradation with different electron acceptors or CH_4-fermentation.

O_2-Reduction		
$7.5O_2 + C_6H_6$	→	$6CO_2 + 3H_2O$
Mn(IV)-Reduction		
$15MnO_2(s) + C_6H_6 + 30H^+$	→	$6CO_2 + 18H_2O + 15Mn^{2+}$
NO_3^--Reactions		
$6NO_3^- + C_6H_6 + 6H^+$	→	$6CO_2 + 6H_2O + 3N_2$
$3.75NO_3^- + C_6H_6 + 0.75H_2O + 7.5H^+$	→	$6CO_2 + 3.75\ NH_4^+$
Fe(III)-Reduction		
$30Fe(OH)_3(s) + C_6H_6 + 60\ H^+$	→	$6CO_2 + 78H_2O + 30Fe^{2+}$
SO_4^{2-}-Reduction		
$3.75SO_4^{2-} + C_6H_6$	→	$6CO_2 + 3H_2O + 3.75S^{2-}$
CH_4-Fermentation		
$4.5H_2O + C_6H_6$	→	$2.25\ CO_2 + 3.75CH_4$

hydrocarbons (e.g. benzene: 1770 mg/L). Redox reactions involved in the degradation of aromatic hydrocarbons (Table 14.3) considerably reduce the further spreading of these hydrocarbon pollutants in aquifers. Table 14.3 shows an overview of the reactions involved in the degradation of benzene, exemplary representing the group of BTEX (benzene, toluene, ethylbenzene and xylenes) aromatic hydrocarbons.

At a former gasworks in Düsseldorf (Germany) where benzene was found to be the major component in BTEX pollution, sulfate reduction prevented the further spreading of the plume in the groundwater (Figure 14.3). Figure 14.2 shows the distribution of the aromatic hydrocarbon concentration in the groundwater before the onset of restoration. At the site of investigation, a naturally occurring (intrinsic) biodegradation of BTEX is ascertained by the decrease of dissolved oxidants (O_2, NO_3^-, SO_4^{2-}; see Figure 14.3), an increased concentration of degradation products in water (CO_2, Fe^{2+}, Mn^{2+}, H_2S-species; see WISOTZKY & ECKERT, 1997) and solids (Sulfide-S and TIC, see Figure 14.4).

Figure 14.3 shows the progression of the BTEX-C and sulfate concentration along the central flow line (position see Figure 14.2). Because of the duration of the pollution (>30 years), a steady state of the plume is reached. A strong increase of the BTEX-C concentration in the source area (about 70 m flow distance in Figure 14.3) is coupled with the decrease of sulfate from 175 mg/L to about 50 mg/L (Figure 14.3). Without this sulfate-dominated biodegradation, a higher BTEX concentration downgradient and a larger plume would be the result. Dispersion causes an increase in sulfate concentration further downgrading without reaching the sulfate concentration upgradient of the contamination source. Sulfide, as a product of sulfate reduction reaches only a maximum concentration of 300 µg/L. Therefore, a formation of sulfide minerals is likely to occur. Hydrochemical calculations indicate a saturation with regard to an iron sulfide mineral

(FeS, Mackinawite) and a subsaturation with regard to ferrihydrite in the polluted region (see WISOTZKY & ECKERT, 1997).

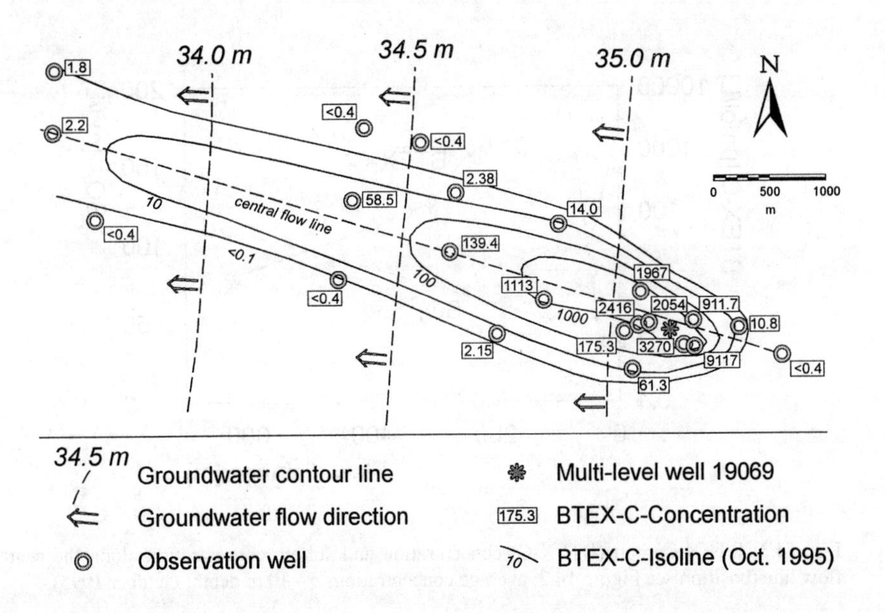

Fig. 14.2: Distribution of BTEX-carbon (BTEX-C) concentration (isolines; in [μmol/L]), position of the central flow line and core drillings (near Multi-level well 16069) as well as the position of the multi-level observation well 19069 (average values 5-10 m depth; October 1995).

Aside from these saturation index calculations, the sulfide-sulfur content of the sediments (Figure 14.4) confirms the formation of sulfide minerals in the plume. Core drillings (n = 11) were applied 20 m down-gradient of the source area (position see Figure 14.2). The results indicate a strong variability of the sulfide-sulfur and TIC-content (total inorganic carbon) in about 100 investigated sediment samples (not shown). However, the depth-specific average of measured concentrations shows a very clear depth-specific distribution of sulfide-sulfur and inorganic carbon (TIC; Figure 14.4). Despite a strong inhomogeneity, a distinct maximum of sulfide-sulfur contents was measured in a depth of 6 to 8 m and below the BTEX-polluted groundwater zone (Figure 14.4). The maximum average sulfide-sulfur content with a value of 0.02 wt% was measured at the lower margin of the BTEX pollution plume (7.5 - 8.5 m depth; Figure 14.4). Therefore, the oxidative BTEX degradation mainly occurs in the mixing zone of reduced carbon (BTEX) and the oxidant (mainly sulfate). Generally, the contact made with the oxidants mainly at the boundaries prevents the pollution from spreading further.

Carbon dioxide released by BTEX degradation contributes to the formation of carbonate minerals in the sediments. Therefore, a maximum TIC content with an average of 0.17 wt% was detected at the lower margin of the pollution plume

(7.5 - 8.5 m depth in Figure 14.4). This TIC distribution indicates a secondary formation of carbonate minerals (measured as inorganic carbon) caused by degradation reactions.

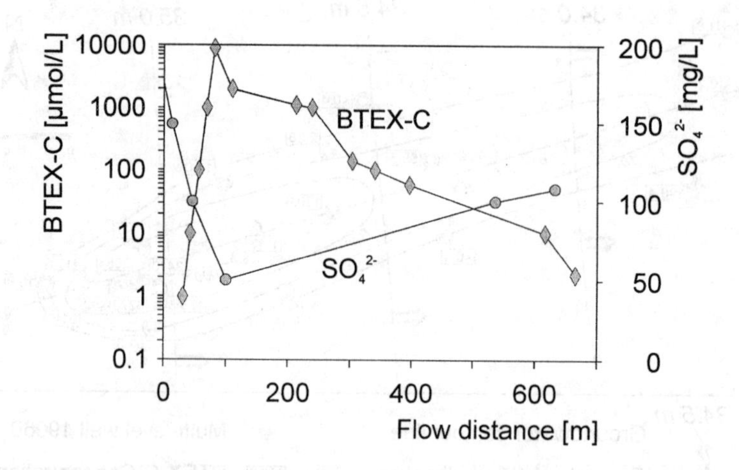

Fig. 14.3: BTEX-Carbon (BTEX-C) concentration and sulfate concentration along the central flow line (position see Figure 14.2; average concentrations 5 - 10 m depth; October 1995).

Ferrous-iron concentrations reach minimum values around 2 mg/L in the polluted depth zone in which the sulfide concentrations are at the maximum. This opposite behaviour of ferrous-iron and dissolved sulfide indicate a solubility limitation (e.g. FeS; see WISOTZKY & ECKERT, 1997).

Besides the changes in BTEX and sulfate concentrations, Figure 14.4a shows the distribution of sulfate isotopes ($\delta^{18}O\text{-}SO_4^{2-}$ and $\delta^{34}S\text{-}SO_4^{2-}$) with increasing depth. The $\delta^{18}O\text{-}SO_4^{2-}$- and $\delta^{34}S\text{-}SO_4^{2-}$-values measured in the sulfate reduction zone (6-9 m depth) were distinctly higher than those of the deeper zone (≥10 m depth). The $\delta^{34}S\text{-}SO_4^{2-}$-values increase from +6 ‰ in the unpolluted zone (≥10 m depth) to a maximum value of +41 ‰ in 7 m depth. The $\delta^{18}O\text{-}SO_4^{2-}$-values show a comparable pattern of distribution. The $\delta^{18}O\text{-}SO_4^{2-}$-values increase from about +6 ‰ in the unpolluted zone (≥10 m depth) to a maximum value of +15 ‰ in the polluted zone (7 m depth). Therefore, the strong sulfate reduction is also confirmed by the accumulation of heavier isotopes (^{34}S and ^{18}O) in the remaining dissolved sulfate pool.

Negative redox potentials (pε = -0.5 to -1.6) are measured at all depths and a minimum value of pε = -1.6 is reached at a depth of 8 m (not shown). The redox potential values measured in the BTEX-polluted groundwater investigated differ from those in the dump aquifer (Table 14.2). In order to test the plausibility of redox measurements in environments with confirmed sulfate reduction, a stability diagram was calculated with the newly developed computer code based on

PHREEQM. The water analysis used in the calculation (observation well 19069, 7 m) was taken from the centre of the pollution area (position see Figure 14.2).

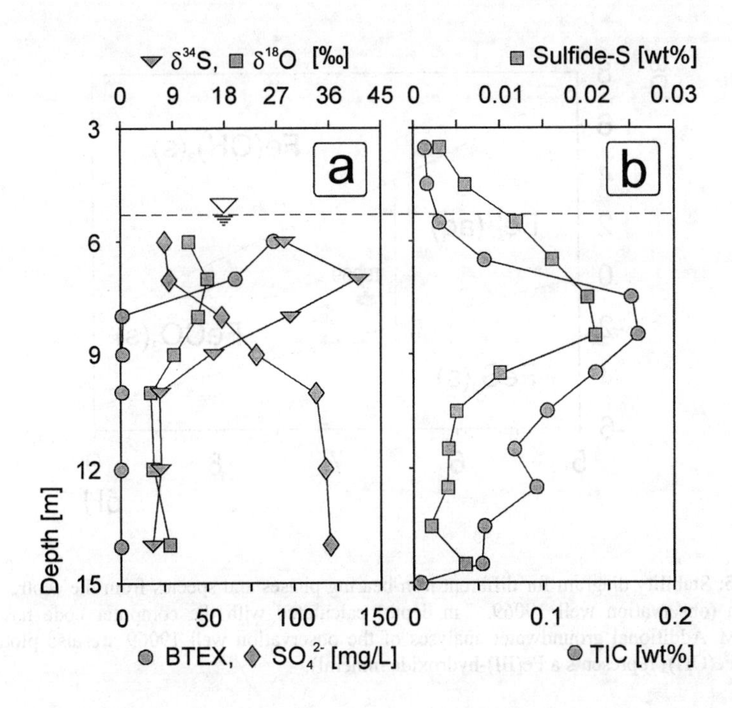

Fig. 14.4: Depth-specific distribution of BTEX concentration, sulfate concentration and the isotopic values of $\delta^{18}O\text{-}SO_4^{2-}$ and $\delta^{34}S\text{-}SO_4^{2-}$ at the multi-level well 19069 (a), and of the average sulfide-sulfur (sulfide-S) and total inorganic carbon content TIC (b) of the sediments.

Figure 14.5 shows that the pH-pε-values of the groundwater analyses are located near the stability boundary of pyrite. The distance between measured values and the stability field of ferrihydrite demonstrates the probability of reductive dissolution of this solid phase (or of Fe(III)-oxides/hydroxides). The released ferrous-iron also leads, other than an increase of dissolved iron, to the formation of iron sulfide minerals in the sediments (see Figure 14.4). Increased TIC-values (see Figure 14.4) show the formation of carbonate minerals as related to the oxidative degradation of BTEX. In the stability diagram (Figure 14.5), siderite is calculated as the stable carbonate mineral phase.

The pε-values measured in this environment (characterised by sulfate reduction) have a plausible order of magnitude with regard to the precipitating phases (probably iron sulfides and siderite), but not to the dissolving phases (probably Fe(III)-hydroxides). Thus, the results only show that a partial redox equilibrium exists in the polluted aquifer zone. Despite the lack of a complete redox equilibrium (see also LINDBERG & RUNNELS, 1984; STUMM & MORGAN, 1996), E_H-

measurements are useful (see pH-pε-diagrams) in hydrogeochemical studies of comparable environments. A complete redox equilibrium cannot be expected because an active system develops from disequilibrium to equilibrium.

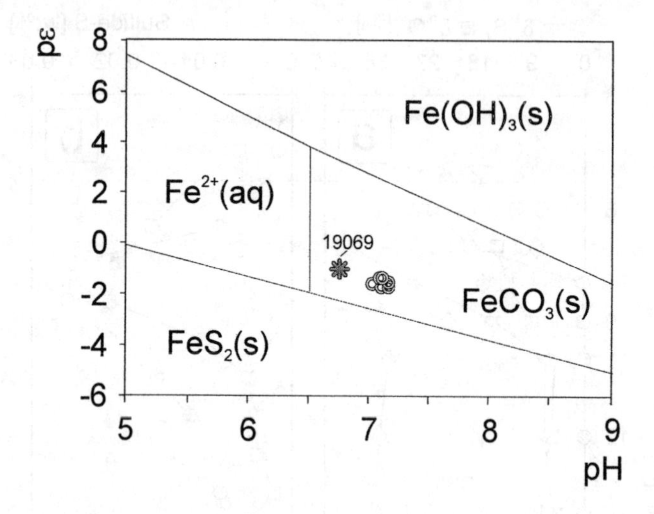

Fig. 14.5: Stability diagram for different iron-bearing phases and species from the centre of the pollution (observation well 19069, 7 m depth) calculated with the computer code based on PHREEQM. Additional groundwater analyses of the observation well 19069 are also plotted as circles ($Fe(OH)_3$ represents a Fe(III)-hydroxide mineral).

The precipitated sulfide minerals are only stable at certain E_H-values and when contact is made with a degradable groundwater pollution. Re-oxidation of the sulfide minerals probably occurs when the BTEX pollution is reduced. This finding can be explained with an input of oxygen or nitrate with the groundwater recharge into the sulfidic environment. Therefore, the formation of sulfide minerals in a BTEX-polluted aquifer is a temporal effect, however, it limits the spreading of the pollution plume.

14.4 Summary

The redox potential measured in the investigated lignite dump aquifer is sensitive to the absence (in the groundwater) or presence (natural groundwater discharge) of oxygen or the change of the redox state of the dissolved iron (Fe(II)/Fe(III). The redox potentials seem to be determined by the Fe^{2+}/Fe^{3+} and the $Fe^{2+}/Fe(OH)_3(s)$ redox couples. But the remaining primary pyrite in the sediments (not oxidised during mining activity) does not seem to have an effect on the measured redox values of groundwater. Thus, complete redox equilibrium is not reached in the overburden dump aquifer.

The sulfate reduction in a BTEX pollution zone, mainly characterised by the presence of benzene, limits the further spreading of the plume in the groundwater at the former gasworks in Düsseldorf (Germany). Despite strong inhomogeneities, sulfide-sulfur in the solid phase was measurable in the BTEX-polluted groundwater zone and below. The maximum sulfide-sulfur content (average values) was measured at the bottom of the BTEX pollution plume. Therefore, the oxidative BTEX degradation mainly takes place in the mixing zone of reduced carbon (BTEX) and the oxidant (mainly sulfate). Generally, contact with oxidants at the boundaries limits the further spreading of the pollution. Carbon dioxide released BTEX degradation leads to the formation of carbonate minerals in the sediments (enhanced concentration of TIC). The $\delta^{18}O\text{-}SO_4^{2-}$- and $\delta^{34}S\text{-}SO_4^{2-}$-values measured in the sulfate reduction zone are distinctly higher than those in the deeper zone. The $\delta^{34}S\text{-}SO_4^{2-}$-values increase from +6‰ in the unpolluted zone to a maximum value of +41‰ in the polluted zone. The $\delta^{18}O\text{-}SO_4^{2-}$-values show a comparable distribution pattern: The $\delta^{18}O\text{-}SO_4^{2-}$-values increase from about +6‰ in the unpolluted zone to a maximum value of +15‰ in the polluted zone. Therefore, the strong sulfate reduction is also confirmed by the accumulation of heavier isotopes (^{34}S and ^{18}O) in the remaining dissolved sulfate pool. The pε-values measured in this environment assume a plausible order of magnitude with regard to the precipitating phases (probably iron sulfides and siderite), but not to the dissolving phases (probably Fe(III)-hydroxide minerals). As a result, there is only a partial redox equilibrium in the BTEX-polluted aquifer zone as well.

14.5 References

ALPERS, C.N. & BLOWES, D.W. (1994): Environmental Geochemistry of Sulfide Oxidation. Am. Chem. Soc., Washington DC.

APPELO, C.A.J. & WILLEMSEN, A. (1987): Geochemical Calculations and Observations on Salt Water Intrusions, I - A Combined Geochemical/Mixing Cell Model. J. Hydrol.: 94: 313-330.

BARBARO, J.R.; BARKER, J.F.; LEMON, L.A. & MAYFIELD, C.I. (1992): Biotransformation of BTEX under Anaerobic, Denitrifying Conditions: Field and Laboratory Observations. J. Cont. Hydrology. 11: 245-272.

BARKER, J.F.; PATRICK, G.C. & MAJOR, D. (1987): Natural Attenuation of Aromatic Hydrocarbons in a Shallow Sand Aquifer. Ground Water Monitoring Review 7,1: 64-71.

BIGHAM, J.M.; SCHWERTMANN, U.; TRAINA, S.J.; WINLAND, R.L. & WOLF, M. (1996): Schwertmannite and the Chemical Modeling of Iron in Acid Sulfate Waters. Geochim. Cosmochim. Acta. 60: 2111-2121.

BROOKINS, D.G. (1987): Eh-pH Diagrams for Geochemistry. Springer, Berlin.

BRUMSACK, H.-J. (1981): A Simple Method for the Determination of Sulfide- and Sulfate-Sulfur in Geological Materials by Using Different Temperatures of Decomposition. Z. Anal. Chem. 307: 206-207.

CHAMBERS, L.A. & TRUDINGER, P.A. (1979): Microbiological Fractionation of Stable Sulfur Isotopes: A Review and Critique. Geomicrobiology Journal 1: 249-293.

CORNELL, R.M. & SCHWERTMANN, U. (1996): The Iron Oxides - Structure, Properties, Reactions, Occurrence and Uses. VCH Verlagsgesellschaft mbH, Weinheim.

FRIESE, K.; HUPFER, M. & SCHULTZE, M. (1998): Chemical Characteristics of Water and Sediment in Acid Mining Lakes of the Lusatian Lignite District. In: GELLER, W.; KLAPPER, H. & SALOMONS, W. (Eds.): Acidic Mining Lakes. Springer, Berlin, pp 25-45.

FROELICH, P.N.; KLINKHAMMER, G.P.; BENDER, M.L.; LUEDTKE, N.A.; HEATH, G.R.; CULLEN, D.; DAUPHIN, P.; HAMMOND, D.; HARTMANN, B. & MAYNARD, V. (1979): Early Oxidation of Organic Matter in Pelagic Sediments of the Eastern Equatorial Atlantic: Suboxic Diagenesis. Geochim. Cosmochim. Acta 43: 1075-1090.

GARRELS, R.M. & CHRIST, C.L. (1965): Solutions, Minerals, and Equilibria. Harper and Row, New York

HOFFMANN, K. (1993): Altlastenproblematik auf ehemaligen Zechen- und Kokereistandorten. In: WIGGERING, H. (Ed.): Steinkohlenbergbau - Steinkohle als Grundstoff, Energieträger und Umweltfaktor. Ernst und Sohn, Berlin, pp 186-203.

LEUCHS, W. (1988): Vorkommen, Abfolge und Auswirkungen anoxischer Redoxreaktionen in einem pleistozänen Porengrundwasserleiter. Besondere Mitteilungen zum Dtsch. Gewässerkdl. Jb. 52: 106 p.

LINDBERG, R.D. & RUNNELS, D.D. (1984): Ground Water Redox Reactions: An Analysis of Equilibrium State Applied to Eh Measurements and Geochemical Modelling. Science 225: 925-927

MAJOR, D.W.; MAYFIELD, C.I. & BARKER, J.F. (1988): Biotransformation of Benzene by Denitrification in Aquifer Sand. Ground Water 26: 8-14.

NORDSTROM, D.K.; JENNE, E.A. & BALL, J.W. (1979): Redox Equilibria of Iron in Acid Mine Waters. In Jenne EA [Ed.] Chemical Modelling in Aqueous Systems. Am. Chem. Soc. Symp. Ser. 93: 51-79.

PATTERSON, B.M.; PRIBAC, F.; BARBER, C.; DAVIS, G.B. & GIBBS, R. (1993): Biodegradation and Retardation of PCE and BTEX Compounds in Aquifer Material from Western Australia Using Large-Scale Columns. J. Cont. Hydrology 14: 261-278.

PIETSCH, W. (1979): Zur hydrochemischen Situation der Tagebauseen des Lausitzer Braunkohlen-Reviers. Arch. Naturschutz Landschaftsforsch. 19: 97-115.

SCHMITT, R.; LANGGUTH, H.R.; PÜTTMANN, W.; ROHNS, H.P.; ECKERT, P. & SCHUBERT, J. (1996): Biodegradation of Aromatic Hydrocarbons under Anoxic Conditions in a Shallow Sand and Gravel Aquifer of the Lower Rhine Valley, Germany. Org. Geochem. 25: 41-50.

SIGG, L. & STUMM, W. (1994): Aquatische Chemie. Verlag der Fachvereine Zürich, Zürich

STUMM, W. & MORGAN, J.J. (1996): Aquatic Chemistry - Chemical Equilibria and Rates in Natural Waters. John Wiley & Sons, New York.

VAN BERK, W. & WISOTZKY, F. (1995): Sulfide Oxidation in Brown Coal Overburden and Chemical Modelling of Reactions in Aquifers Influenced by Sulfide Oxidation. Environmental Geology 26: 192-196.

WISOTZKY, F. (1994): Untersuchungen zur Pyritoxidation in Sedimenten des Rheinischen Braunkohlenreviers und deren Auswirkungen auf die Chemie des Grundwassers. Besondere Mitteilungen zum Dtsch. Gewässerkl. Jb. 58: p 153.

WISOTZKY, F. & ECKERT, P.(1997): Sulfat-dominierter BTEX-Abbau im Grundwasser eines ehemaligen Gaswerksstandortes. Grundwasser 2,1: 11-21.

WISOTZKY, F. (1998a): Chemical Reactions in Aquifers Influenced by Sulfide Oxidation and in Sulfide Oxidation Zones. In: GELLER, W.; KLAPPER, H. & SALOMONS, W. (Eds.): Acidic Mining Lakes. Springer, Berlin, pp 223-236.

WISOTZKY, F. (1998b): Hydrogeochemische Modellierung von abbauwirksamen Redoxreaktionen bei einer BTEX-Belastung des Grundwassers. In: GRAMBOW, B. & FANGHÄNEL, T. (Eds.): Geochemische Modellierung - Radiotoxische und chemisch-toxische Stoffe in natürlichen aquatischen Systemen. Forschungszentrum Karlsruhe, Wissenschaftliche Berichte FZKA 6051, pp 149-160.

WISOTZKY, F. (1998c): Assessment of the Extent of Sulfate Reduction in Lignite Mining Dumps using Thermodynamic Equilibrium Models. Water, Air, and Soil Pollution 108: 285-296.

Chapter 15

Redox Buffer Capacity Concept as a Tool for the Assessment of Long-Term Effects in Natural Attenuation / Intrinsic Remediation

F. von der Kammer, J. Thöming & U. Förstner

15.1 Introduction

15.1.1 Natural Attenuation

The conceptual approach of natural attenuation/intrinsic remediation, which relies on natural subsurface processes rather than traditional engineered procedures to eliminate contaminants in groundwater or soil, can provide tailored measures for low-risk cases, such as contamination by petroleum hydrocarbons. In this concept, increased efforts will be required both for initial studies on site hydrology, geochemistry and microbiology as well as subsequent monitoring of the remediation success.

The concept of natural attenuation is based on processes such as biological degradation, dispersion, dilution, sorption, evaporation and/or chemical and biochemical stabilisation of pollutants. Toxicity, mobility or volume of pollutants should be reduced to an extent, that human health and ecosystems are no more endangered (NYER & DUFFIN, 1997).

Major advantages of the concept are – as for most *in situ* procedures – the avoidance of secondary wastes and reduction of hazards for exposed persons compared to *ex situ* treatments. Problems may arise from the long time for reaching the remediation goals including subsequent surveillance. In the opinion of U.S. Environmental Protection Agency, intensive instruction will be needed to receive public acceptance of this concept (EPA 1997).

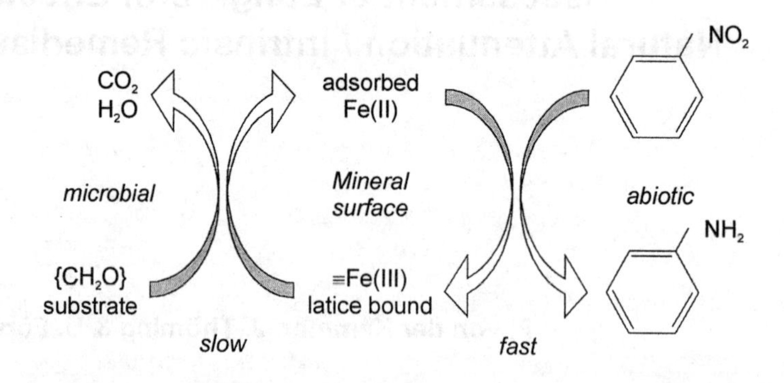

Fig. 15.1: Reaction scheme postulating the occurrence of microbial and abiotic reactions involved in the reduction of NACs in laboratory aquifer columns (KLAUSEN et al., 1995).

Typical for natural attenuation strategies is the implementation of biological, chemical and geotechnical approaches. Common objectives are: site characterisation with regard to the efficiency of the expected retardation/degradation mechanisms, proof of applicability of the natural attenuation concept (i.e., time frame) and elucidation of questions relating to the persistence of critical pollution sources.

15.1.1.1
Redox Processes in the Natural Attenuation Concept

At this stage, three priority themes have been identified for research and development in the field of natural attenuation/intrinsic remediation (FÖRSTNER, 1998):

- *Reconnaissance and monitoring*: Concepts, sampling strategies and analytical procedures;
- Long-term behaviour of organic contaminants in leachates and groundwater. Major aspects are "degradation budgets", "rest-risk" and "reliable prognosis";
- Natural attenuation at large-scale deteriorations of groundwater by former landfills. Major topic: Effect of redox variations. Apart from traditional aero-

bic degradation, which is conducted by a wide range of micro-organisms (bacteria, actinomycetes, fungi), there are processes of anaerobic decay, where micro-organisms receive their energy either from the utilisation of oxidised compounds like nitrate, sulfide or Fe(III) as electron acceptors or by fermentation of organic substance (fermentation: splitting of organic molecules into oxidised and reduced fragments).

Laboratory and field studies have indicated that the reduction of organic pollutants may also involve abiotic chemical reactions. Given the abundance and the range of reduction potentials of Fe(II) species that may exist in anoxic environments, it seems likely that, particularly under iron-reducing conditions, such species play a pivotal role as electron donors or electron transfer mediators in redox transformations of organic compounds. As an example, Figure 15.1 suggests that the reduction of nitro-aromatic compounds (NACs) occurs by oxidation of surface-bound Fe(II) that has been continuously produced by dissimilatory iron-reducing micro-organisms and that the (re)generation of these active Fe(III) surface species (and not the actual electron transfer to the NACs) is rate limiting (KLAUSEN et al., 1995).

15.1.1.2
Practical Experience and Priority Parameters

There is already considerable experience with the redox theme in the field of contaminated soil and groundwater. An anaerobic field injection experiment was performed in a landfill leachate plume (RÜGGE et al., 1998). The field injection lasted for 195 days, and the injected cloud was monitored over a period of 924 days as it travelled through a redox environment dominated by iron reduction. Anaerobic degradation was observed for toluene, o-xylene, naphthalene, while benzene was not degraded in the aquifer.

Such differentiated degradation processes are even more pronounced for chlorinated ethenes. For example, comprehensive laboratory and field studies indicate, that tetra- and trichloroethylene is easily degraded in the methanogenic and sulfate-reducing area of the plume, whereas dechlorination of the follow-up compound vinyl chloride does not occur, as long as at least low concentrations of dissolved organic carbon (DOC) are available (HOLMES et al., 1998).

Important milieu parameters for estimating natural attenuation are the concentration of oxygen, nitrate, iron(II), sulfate, methane, and – less significant – manganese, calcium, bicarbonate and pH. Interpretation of such data can rely on combinations of *in situ*-microcosms, laboratory batch studies and field observations (CHRISTENSEN et al., 1994). More advanced methodological concepts include geochemical speciation models, mainly for iron (oxide, sulfide, carbonate, Fe-complexes.

A recent proposal for standardised testing of applicability of the natural attenuation approach has been based on the three criteria:

1. Extent of reductive dehalogenation,
2. redox value, and
3. DOC (RIJNAARTS et al., 1998; TONNAER et al., 1998).

15.1.2 Contaminant Plumes in Groundwater

Redox processes in the subsurface play a major role for the mobility, transport and attenuation of contaminants (BARCELONA & HOLM, 1991). The infiltration of pollutants together with dissolved natural organic matter (e.g.: leachate from unsecured landfills) or biodegradable organic contaminants (e.g.: from airfield fuel storage) into shallow aerobic aquifers usually initiate the development of redox successions (BARCELONA et al., 1989; LYNGKILDE & CHRISTENSEN, 1992b). These successions occur in a reverse order compared to deep pristine aquifers (LOVELEY et al., 1994). Downgradient transport of contaminants is therefore often accompanied by the development of reducing zones in the groundwater.

Excessive reduction processes close to the source of biodegradable carbon form a methanogenic zone, due to a depletion of available oxidants. Downgradient of the source, the redox potential approaches the former oxic conditions through a succession of more or less defined redox conditions (Figure 15.2).

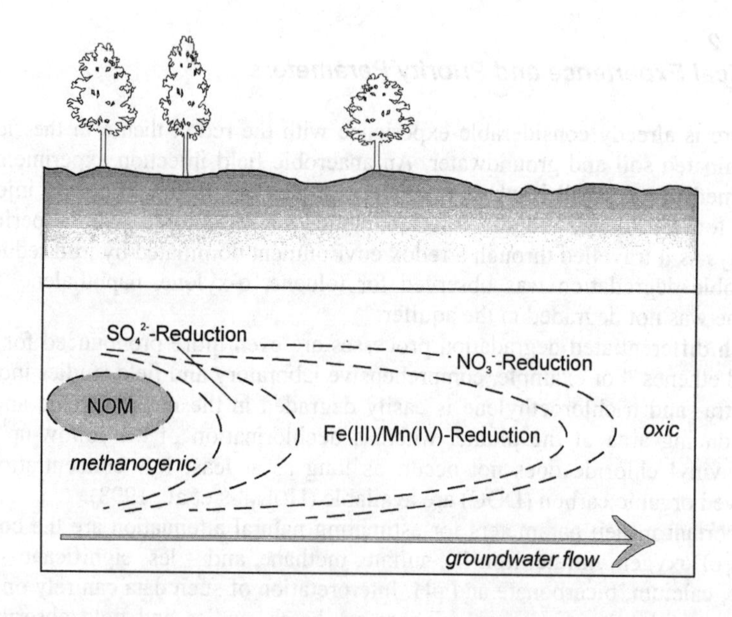

Fig. 15.2: Schematic plot of a reverse redox succession in a contaminated aquifer and the related terminal electron-accepting processes (TEAP).

The individual zones can be labelled by the occurring terminal electron accepting processes (TEAP). The actual sequence of zones is depending on the available chemical species contributing to the redox processes. The spatial extension of particular zones and the whole reducing plume depends on the oxidation capacity (OXC) of the aquifer compared to the amount of reactive reductants which are introduced into the aquifer over time.

15.1.3 Redox Buffer Capacity

The general approach for the assessment of the redox status of aquifers are platinum electrode measurements or chemical equilibrium calculations based on the chemical analysis of the main redox species (BARCELONA et al., 1989; LINDBERG & RUNNELS, 1984; HOSTETTLER, 1984). Beside the widely discussed problems of measuring redox potentials in natural waters (LINDBERG & RUNNELS, 1984), additional problems occur in contaminated aquifers due to high reactivity, disequilibrium and the distribution of reduced species (e.g. CH_4, $Fe(II)_{soluble}$) into the more oxic parts of the plume (LOVLEY et al., 1994). In fact, meaningful interpretation of redox potentials, acquired by electrode measurements from partially oxidised soils, seems to be impossible (BARTLETT & JAMES, 1995).

Additionally it has to be pointed out that the measurement of redox potential, as the single pH measurement, is a capture of current conditions. There is little opportunity to assess or predict the reactions within the plume quantitatively or the future behaviour of the system.

Tab. 15.1: TEAP-reactions in aquifers (in the order of decreasing redox-potential) and their estimated contribution to the oxidation capacity (OXC) of the aquifer (derived from HERON & CHRISTENSEN, 1994).

Half-Reaction	*Phase*	*Contribution to OXC*
$O_2 + 4H^+ + 4e^- \rightarrow 2H_2O$	aqueous	low
$2NO_3^- + 12H^+ + 10e^- \rightarrow N_2 + 6H_2O$	aqueous	low
$Mn^{4+} + 2e^- \rightarrow Mn^{2+}$	solid	high
$Fe^{3+} + e^- \rightarrow Fe^{2+}$	solid	high
$SO_4^{2-} + 9H^+ + 8e^- \rightarrow HS^- + 4H_2O$	aqueous + solid	low
$CH_2O + 4e^- + 4H^+ \rightarrow CH_4 + H_2O$	aqueous + solid	high (only potential)

A selection of the potentially occurring electron accepting half-reactions is given in Table 15.1 together with their estimated contribution to the oxidation capacity of the aquifer. The contribution of organic matter (expressed as CH_2O) to the oxidation capacity (OXC) is calculated for a complete reduction from oxidation state 0 (e.g. glucose) to -4 (methane). There is only a potential contribution to OXC, since this reaction needs extremely low redox potentials and it is unknown to what extent organic mater can contribute to OXC under natural conditions (HERON & CHRISTENSEN, 1994).

Despite the uncertainty for the organic matter, HERON & CHRISTENSEN (1994) calculated the oxidation capacity as:

$$OXC = 4[O_2] + 5[NO_3^-] + [Fe(III)] + 2[Mn(IV)] + 8[SO_4^{2-}] + 4[TOC] \tag{15.1}$$

Recent research showed that natural organic matter can be partially and reversibly reduced by micro-organisms and can act as an electron-shuttle (LOVLEY et al., 1998). This shuttle can reduce insoluble Fe(III)-phases or Fe(III) captured within micro-pores that are not directly accessible for the reducing bacteria.

But in general, bio-available organic matter should be the main electron-donor and therefore express the reduction capacity of the liquid phase as long as they occur in solution. Inorganic species can contribute as well , especially if they are soluble and transported downgradient in the plume (e.g.: $Fe(II)_{soluble}$).

For a risk assessment of contaminated sites, or for natural attenuation process management, it is of outstanding importance to know how and to what extent the individual anaerobic parts of the plume will spread or decline in the future. For anaerobic degradation of pollutants (e.g. tetrachloroethylene) or aerobic degradation of follow-up metabolites (vinyl chloride) the extent and dynamic of redox zones under study has to be examined. LYNGKILDE & CHRISTENSEN (1992a) outlined the importance of the methanogenic, sulfidic and iron reducing zone for the attenuation of several organic pollutants. Hence, it is essential to have information about recent and future development of redox conditions in contaminated aquifers.

15.2 Redox Buffer Capacities: Conceptional Approach

Even assuming optimal conditions regarding the analytical equipment used (KÖLLING, 1985), electrode measurements of redox potential in groundwaters can only give a current value. Regardless of the worth of the obtained number, it is impossible to make a prediction of the future development of redox conditions in contaminated, and therefore highly reactive, aquifers.

This problem is well known from the master variable pH and can be similarly treated by introducing a capacity component. While the acid or base neutralisation capacities represent a certain amount of protons that can be accepted or donated until a defined threshold limit (usually pH 4 or 10) is reached, the reduction or oxidation capacities represent the amount of electrons that can be donated or accepted. No threshold limits have been defined yet and it will have to be discussed if it is appropriate and/or possible to define certain E_H-limits. Usually the redox buffer capacities are measured against a certain strong reductant or oxidant to provide a quantitative reaction within short-term laboratory testing.

Oxidation capacity (OXC) is the amount of electrons that can be accepted and TRC (total reduction capacity) is the amount of electrons that can be delivered from a sample. Within an oxic aquifer the solid matrix will have a certain OXC, but TRC will approach zero as long as natural Fe(II) or sorbed organic matter is not captured by the TRC measurement. A partially reduced aquifer matrix will have a certain decreased OXC and also a TRC if Fe(II)-components from reduction processes are present. In strongly reduced aquifers the OXC of the matrix approaches zero (HERON & CHRISTENSEN, 1995).

The TRC of the aquifer matrix consists of reduced components that are sorbed to or into the aquifer matrix structure (mainly Fe(II) and organic matter) or from natural, mineral-bound Fe(II) that is captured by the method. Depending on the mineral composition of the aquifer material oxic aquifers may also have a TRC, due to the existence of natural Fe(II) components (e.g. from magnetite).

Like the base or acid neutralisation capacity, the reduction or oxidation capacity is mainly represented by the solid aquifer matrix, while the driving force or component (reducing reagent, mainly DOC) is introduced in solution. The OXC-TRC concept has to be seen within the environmental context, that means only possible and relevant redox processes should be included.

As it can be seen from Table 15.1, iron and manganese play a major role for the OXC of the matrix. Hence, analytical methods focus on these Fe/Mn-phases.

15.3 Analytical Approach and Application

15.3.1 Overview

Several methods are published for the assessment of the oxidation capacity of aquifer matrix material.

15.3.1.1 OXC

Usually a strong reductant is used in excess and the oxidation capacity (OXC) is calculated from the loss of the reducing reagent. Cr(II) was used by BARCELONA & HOLM (1991) and loss of Cr(II) was measured photometrically. Ti(III)/EDTA was used by HERON et al. (1994) and HERON & CHRISTENSEN (1995). The applied Ti(III)/EDTA-method extracted Fe(III) and Fe(II) components quantitatively from standard minerals with exception of magnetite. Fe_{tot} was measured by AAS, while the OXC was determined by dichromate titration of the remaining Ti(III). The OXC should express the amount of extractable Fe(III) without the contribution of magnetite. Fe(II) can be calculated as the difference of Fe_{tot} and Fe(III).

The Ti(III)/EDTA-extraction was favoured by HERON et al. (1994) after testing several iron extraction techniques. Beside the Ti(III)/EDTA-extraction, acidic extractions with following photometrical ferrozine-based Fe(III)/(II)-speciation (STOOKEY, 1970; LOVLEY & PHILLIPS, 1987) and reductive dissolution methods were examined.

In contrast to what has been mentioned before, LOVLEY & PHILLIPS (1987) developed an acidic hydroxylamine extraction method followed by ferrozine based Fe(III)/Fe(II) speciation. They compared the results with microbially reduced Fe(III) from laboratory incubations of the same samples and concluded that hydroxylamine extracted iron corresponds to the Fe(III) available for microbial reduction.

15.3.1.2
TRC

The reduction capacity (TRC) is obtained by using sodium dichromate(IV) for a total breakdown of organic components and oxidation of reduced species (Fe(II), Mn(II) if present, sulfides) and leads to the total reduction capacity (TRC), methods used are similar to the determination of the chemical oxygen demand. An alternative approach was done by THÖMING & CALMANO (1998). Cer(IV)sulfate was used as an oxidising agent.

15.3.2
Case Study: Simplified Illustration of a Groundwater Plume's Redox Activities by Common Sequential Extraction

A major tool to assess long-term effects is, as explained above, the TRC. Such long-term effects can be illustrated following a reducing groundwater plume. On its way through anoxic zones its TRC decreases. The objective of this simplified case study is to give an idea of these long term effects combining TCR measurements with a common sequential extraction. The principle behind this is, that a soil sample of an oxic zone is brought in contact with liquids of varying redox potentials followed by TRC analysis.

15.3.2.1
Determination of the Total Reduction Capacity (TRC)

The total reduction capacity (TRC) indicates the consumption of the oxidation agent $Ce(SO_4)_2$ in [mol/g] of soil at pH = 2. It was measured for six steps of the sequential extraction for both, the soil and the extracting agent. The TCR of the liquid that results in molar concentration was calculated in [mol/g] of soil, taking the solid/liquid-ratio of the extraction step into consideration.

For the TRC-measurements a 0.1 mol/L $Ce(SO_4)_2$ solution with a pH = 2 was made by means of sulfuric acid (CSS). The liquid samples were directly titrated with the CSS-solution potentiometrically. Soil samples of 1 g each were suspended in 35 mL of the CSS-solution for one hour. After filtration, the solution was back-titrated potentiometrically using $FeCl_2$. Hence, the TRC amounts to the difference between the initial and the back-titrated amount of $Ce(SO_4)_2$ per gram of soil. The titrations that were carried out in duplicate analysis showed a relative standard deviation (RSD) of 1.8% (liquids) and 1.9% (solids).

In order to calculate an overall balance for each step, the soil as well as the liquid were measured before and after each step (Figure 15.3). The significance of

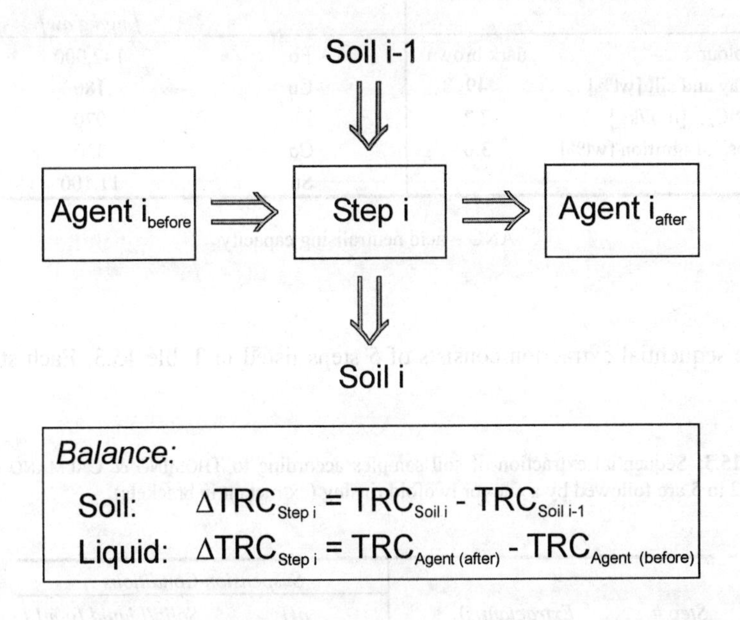

Fig. 15.3: Principle of the overall balances for each step i resulting in ΔTCR values that indicate the TCR shift of each phase during the extraction step.

the data was tested by means of a t-test. Its results, the confidence intervals, were calculated to the total process variance of each series of measurements.

15.3.2.2
Soil Sample and Extraction Scheme

The used sample consisted of the first 10 cm of soil taken from a highly contaminated site. It was contaminated by deposited sediment which was excavated from an effluent drain of an accumulator plant.

In Table 15.2 the characteristics of this soil are given. The obtained 10 kg sample was sieved on site to less than 2 mm and air-dried. Acid neutralisation capacity (ANC_{24}) was measured using a pH stat test. The test maintains a pH = 4 of a 1/10 solid/liquid-ratio soil suspension for 24 hours and records the nitric acid consumption. According to DIN 38414 T3, the loss of ignition (550°C) was determined to estimate the amount of organic matter in the soil. Pseudo total contents of heavy metals within the samples were determined as described in DIN 38414 T7.

Tab. 15.2: Characteristics of the used soil samples.

	Soil sample	*Contaminants*	*Pseudo total content [mg/kg dw]*
Colour	dark brown	Pb	142,000
Clay and silt [wt%]	49	Cu	180
ANC_{24}^{*} [mol/kg]	2.2	Zn	970
Loss of ignition [wt%]	3.6	Cd	330
		Sb	11,100

* ANC = acid neutralising capacity

The sequential extraction consists of 5 steps listed in Table 15.3. Each step of

Tab. 15.3: Sequential extraction of soil samples according to THÖMING & CALMANO (1998). Steps 2 to 5 are followed by a one- or twofold rinsing (extractant in brackets).

		Suspension Conditions	
Step #	*Extractant(s)*	*pH*	*Solid/Liquid [g/mL]*
1	1.0 mol/L NH_4NO_3	≈ 7	2/50
2	1.0 mol/L NH_4OAc (1.0 mol/L NH_4NO_3)	6	2/50
3	0.1 mol/L NH_2OH-HCl + 1.0 mol/L NH_4OAc (1.0 mol/L NH_4OAc)	6	2/50 (twofold)
4	0.025 mol/L NH_4-EDTA (1.0 mol/L NH_4OAc)	4.6	2/50
5	0.2 mol/L NH_4-oxalate (0.2 mol/L NH_4-oxalate)	3.3	2/50

the sequential extraction except the first is followed by a one- or twofold rinsing (agent used is mentioned in brackets) carried out at the same pH as the main step.

15.3.2.3 Results and Discussion of the Case Study

The results of the TRC measurements are shown in Table 15.4. If these data are used as a basis for the overall balances as described above the Figure 15.4 is obtained. It shows the results of these balances for each single step of the sequential extraction. The error bars that illustrate the 95% confidence interval (t-test) do not

Tab. 15.4: Results of the TRC-measurements [µmol/g soil].

Step i	*Solid*	*Extractant*	*Extract*
Initial	720		
1	710	3.4	3
2	715	1.2	3.1
3	690	254	294
4	600	84	91
5	535	929	965

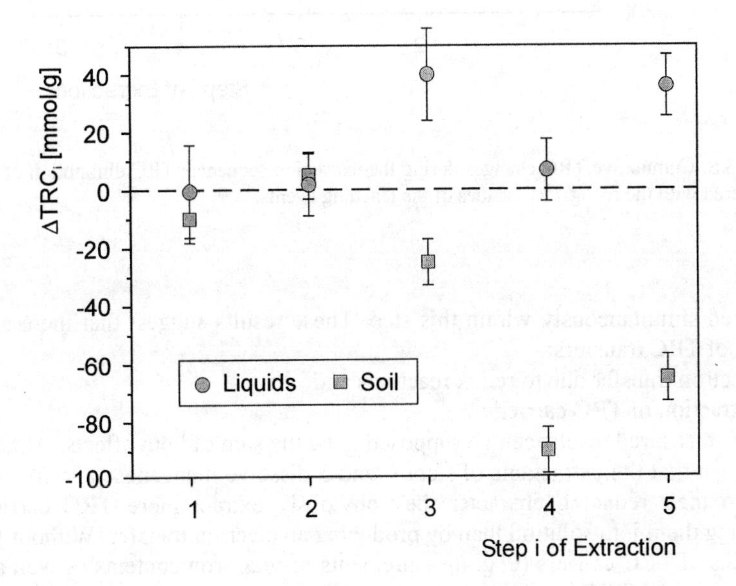

Fig. 15.4: Overall TRC balances for each single step of the sequential extraction. The error bars illustrate the 95% confidence interval (t-test).

overlap for the steps 3, 4 and 5 neither within each series (liquids or soil) nor in-between them. This indicates significant distinctions of the data allowing their comparison. The easiest way of such a comparison is by assessing the cumulative changes as shown in Figure 15.5.

In all steps with significant TRC changes (steps 3, 4 and 5) the absolute TRC of the soil is lowered, while the TRC-values of the liquids are increasing. It seems to be surprising that even the extractant with the lowest redox potential and a TRC-value higher than the TRC of the soil (step 5) increases in its TRC during the extraction procedure. On the other hand, it is typical that, the TRC of the soil is

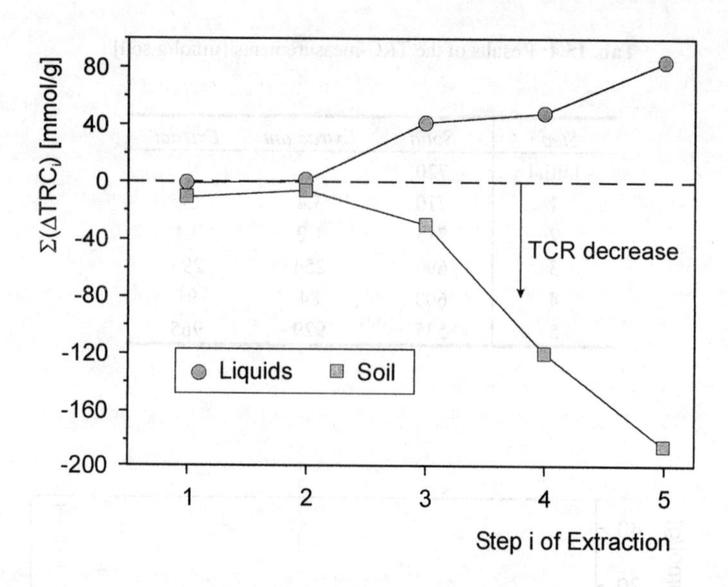

Fig. 15.5: Cumulative TRC changes during the extraction sequence: TRC diminution of the soil compared with the rising TRC-values of the leaching agents.

lowered simultaneously within this step. These results suggest that there are two kinds of TRC transfers:

1. electron transfer due to redox reactions and
2. extraction of TRC carriers.

The measured results can be supposed to be the sum of both effects. Although it is known that the extractants of steps 3 and 5 dissolve manganese and iron oxides due to their reducing character they obviously extract more TRC carriers by bringing them into solution than by producing an electron transfer. Without further analysis of TRC carriers (e. g. measurements of total iron contents as well as iron speciation and TOC measurements) it cannot be distinguished between these two kinds of TRC transfers.

15.4 Conclusions

Natural attenuation is a feasible approach for low-cost cleanup of low-risk contamination. Subsurface redox potential and redox processes are of outstanding importance within this concept. The rise and decay of redox zones in a contaminated aquifer depends on the ratio of electron donators introduced into the aquifer to the electron consumption capacity of the aquifer, the OXC.

In general, the described approach extends the master variable E_H for the essential capacity component and makes the assessment and prediction of long-term redox processes possible. Yet some questions remain.

The analytical methods described are returning results of an operational character. As the case study shows, pure chemical techniques do not give any details about the electron/species transfer. Especially the TRC methods do not consider the actual availability of organic compounds, since the kinetics of biodegradation and biochemical oxidation of Fe(II) compounds are simply disregarded by the use of an extremely strong oxidising agent.

The results of the case study can also be interpreted as a non-destructive transfer of natural organic matter associated with the extracted reductive iron-phases. The transfer of the organic compounds, possessing an unknown TRC, from the solid phase into the extractant of the sequential extraction step would successively decrease the TRC of the matrix and increase the TRC of the extractant.

Further research has to be done on the described methodologies, taking into account the bio-availability, kinetic aspects and the optimal adaptation of laboratory test methods to the occurring reactions in the groundwater plume.

It should not be overseen that introducing a capacity component into the activity parameter E_H will yield extended knowledge about general geochemical processes as well as a powerful tool for the assessment of long term-effects within contaminated aquifers.

15.5 References

BARCELONA, M.J. & HOLM, T.R. (1991): Oxidation-Reduction Capacities of Aquifer Solids. Environ. Sci. Technol. 25: 1565-1572.

BARCELONA, M.J.; HOLM, T.R.; SCHOCK, M.R. & GEORGE, G.K. (1989): Spatial and Temporal Gradients in Aquifer Oxidation-Reduction Conditions. Water Resour. Res. 25: 991-1003.

BARTLETT, R.J. & JAMES, B.R. (1995): System for categorizing soil redox status by chemical field testing. Geoderma 68: 211-218.

CHRISTENSEN, T.H., KJELDSEN, P., ALBRECHTSEN, H.J., HERON, G., NIELSEN, P.H., BJERG, P.L. & HOLM, P.E. (1994): Attenuation of pollutants in landfill leachate polluted aquifers. Crit. Rev. Environ. Sci. Technol. 24: 119-202.

DIN 38414 T3 (1985): Bestimmung des Glührückstandes und des Glühverlustes der Trockenmasse eines Schlammes (S3). Deutsche Einheitsverfahren zur Wasser-, Abwasser- und Schlammuntersuchung. Bd. V, Normenausschuß Wasserwesen im DIN Deutsches Institut für Normung e. V., VCH Verlag, Weinheim.

DIN 38414 T7 (1983): Aufschluß mit Königswasser zur nachfolgenden Bestimmung des säurelöslichen Anteils von Metallen (S7). Deutsche Einheitsverfahren zur Wasser-, Abwasser- und Schlammuntersuchung. Bd. V, Normenausschuß Wasserwesen im DIN Deutsches Institut für Normung e. V., VCH Verlag, Weinheim.

FÖRSTNER, U. (in press): Reaktionen, die einen natürlichen Rückhalt bzw. Abbau von Schadstoffen bewirken. In: FRANZIUS, V. & BACHMANN, G. (Eds.) Bodenschutz & Altlasten 1998; Erich Schmidt Verlag Berlin.

HERON, G.; CROUZET, C; BOURG, A.C.M. & CHRISTENSEN, T.H. (1994): Speciation of Fe(II) and Fe(III) in Contaminated Aquifer Sediments Using Chemical Extraction Techniques. Envir. Sci. Technol. 28: 1698-1705.

HERON, G.; CHRISTENSEN, T.H. & TJELL, J.C. (1994): Oxidation Capacity of Aquifer Sediments. Envir. Sci. Technol. 28: 153-158.

HERON, G. & CHRISTENSEN, T.H. (1995): Impact of Sediment-Bound Iron on the Redox Buffering in a Landfill Leachate Polluted Aquifer (Vejen, Denmark). Envir. Sci. Technol. 29: 187-192.

HOLMES, M.W.; MORGAN, P.; KLECKA, G.M.; KLIER, N.J.; WEST, R.J.; DAVIES, J.W.; ELLIS, D.E.; LUTZ, E.J.; ODAM, J.M.; EI, T.A.; CHAPELLE, F.H.; MAJOR, D.W.; SALVO, J.J. & BELL, M.J. (1998): The natural attenuation of chlorinated ethenes at Dover Air Force Base, Delaware, USA. Contaminated Soil '98/1, 143-152.

HOSTETTLER, J.D. (1984): Electrode Electrons, Aqueous Electrons and Redox Potentials in Natural Waters. Am. J. Sci. 284: 734-759.

KLAUSEN, J.; TRÖBER, S.P.; HADERLEIN, S.B. & SCHWARZENBACH, R. (1995): Reduction of substituted nitrobenzenes by Fe(II) in aqueous mineral suspensions. Environ. Sci. Technol. 29: 2396-2404.

KÖLLING, M. (1985): Vergleich verschiedener Methoden zur Bestimmung des Redoxzustandes natürlicher Wässer. Unpublished Diplom thesis, Universität Kiel.

LINDBERG R.D. & RUNNELS, D.D. (1984): Groundwater Redox Reactions: an Analysis of Equilibrium State Applied to Eh Measurements and Geochemical Modeling. Science 225: 925-927.

LOVLEY, D.R. & PHILLIPS, E.J.P. (1987): Rapid Assay for Microbially Reducible Ferric Iron in Aquatic Sediments. Appl. Environ. Microbiol. 53: 1536-1540.

LOVLEY, D.R.; CHAPELLE, F.H. & WOODWARD, J.C. (1994): Use of Dissolved H_2 Concentrations to Determine Distribution of Microbially Catalyzed Redox Reactions in Anoxic Groundwater. Environ. Sci. Technol. 28: 1205-1210.

LOVLEY, D.R.; FRAGA, J.L.; BLUNT-HARRIS, E.L., HAYES, L.A.; PHILLIPS, E.J.P. & COATES, J.D. (1998): Humic Substances as a Mediator for Microbially Catalyzed Metal Reduction. Acta Hydrochim. Hydrobiol. 26: 152-157.

LYNGKILDE, J. & CHRISTENSEN, T.H. (1992a): Fate of organic contaminants in the redox zones of a landfill leachate pollution plume. J. Cont. Hydr. 10: 291-307.

LYNGKILDE, J. & CHRISTENSEN, T.H. (1992b): Redox zones of a landfill leachate pollution plume (Vejen, Denmark). J. Cont. Hydr. 10: 273-289.

NYER, E.K. & DUFFIN, M.E. (1997): The state of the art of bioremediation. Groundwater Monitoring and Remediation 17(2): 64-69.

RIJNAARTS, H.H.M.; VAN AALST-VAN LEEUWEN, M.A.; VAN HEININGEN, E.; VAN BUYZEN, H.; SINK, A.; VAN LIERE, H.C.; HARKES, M.; BAARTMANS, R.; BOSMA, T.N.P. & DODDEMA, H.J. (1998): Intrinsic and enhanced bioremediation in aquifers contaminated with chlorinated and aromatic hydrocarbons in the Netherlands. Contaminated Soil '98/1, 109-112.

RÜGGE, K.; BJERG, P.L. & CHRISTENSEN, T.H. (1998): Comparison of field and laboratory methods for determination of potential for natural attenuation in a landfill leachate plume (Grindsted, Denmark). Contaminated Soil '98/1, 101-108.

STOOKEY, L.L. (1970): Ferrozine-A New Spectrphotometric Reagent for Iron. Anal. Chem. 42: 779-781.

TONNAER, H.; OTTEN, A.; ALPHENAAR, A. & ROOVERS, C. (1998): Natural attenuation: A basis for developing extensive remediation concepts. Contaminated Soil '98/1, 191-196.

THÖMING, J. & CALMANO W. (submitted): Applicability of Single and Sequential Extractions for Assessing the Potential Mobility of Heavy Metals in Contaminated Soils. Acta hydrochimica et hydrobiologica.

UNITED STATES ENVIRONMENTAL PROTECTION AGENCY, OFFICE OF SOLID WASTE AND EMERGENCY RESPONSE (1997): Use of Monitored Natural Attenuation at Superfund, RCRA Corrective Action, and Underground Storage Tank Sites. Draft Interim Final OSWER Directive 9200.4-17, 25 pp.

Chapter 16

Redox Zones in the Plume of a Previously Operating Gas Plant

K. Weber, N. Brandsch, B. Reichert, M. Eiswirth, H. Hötzl, O. Hümmer & A. Dahmke

16.1 Introduction

Within the framework of a current research program, the fate of BTEX has been investigated in a Quaternary porous aquifer below a former gas plant in south-west Germany. The investigation site is located in the Neckar valley and occupies an area of about 600 m by 400 m (Figure 16.1). Industrial activities in this area started in the year 1875 with the production of gas and other coal-derived oils and tars by hard coal carbonisation. After 1956 gas production was carried out on the basis of heavy oil; starting in 1964, light mineral oil products were used. There was a change from city gas production to the distribution of natural gas between 1969 and 1974.

Soil and groundwater at the investigation site are highly polluted with mono-aromatic hydrocarbons. The contamination occurred from chronic leaks in tanks and spills during the production of benzene. Also, the gas plant was a target for aerial attacks during World War II (HESKE, 1998).

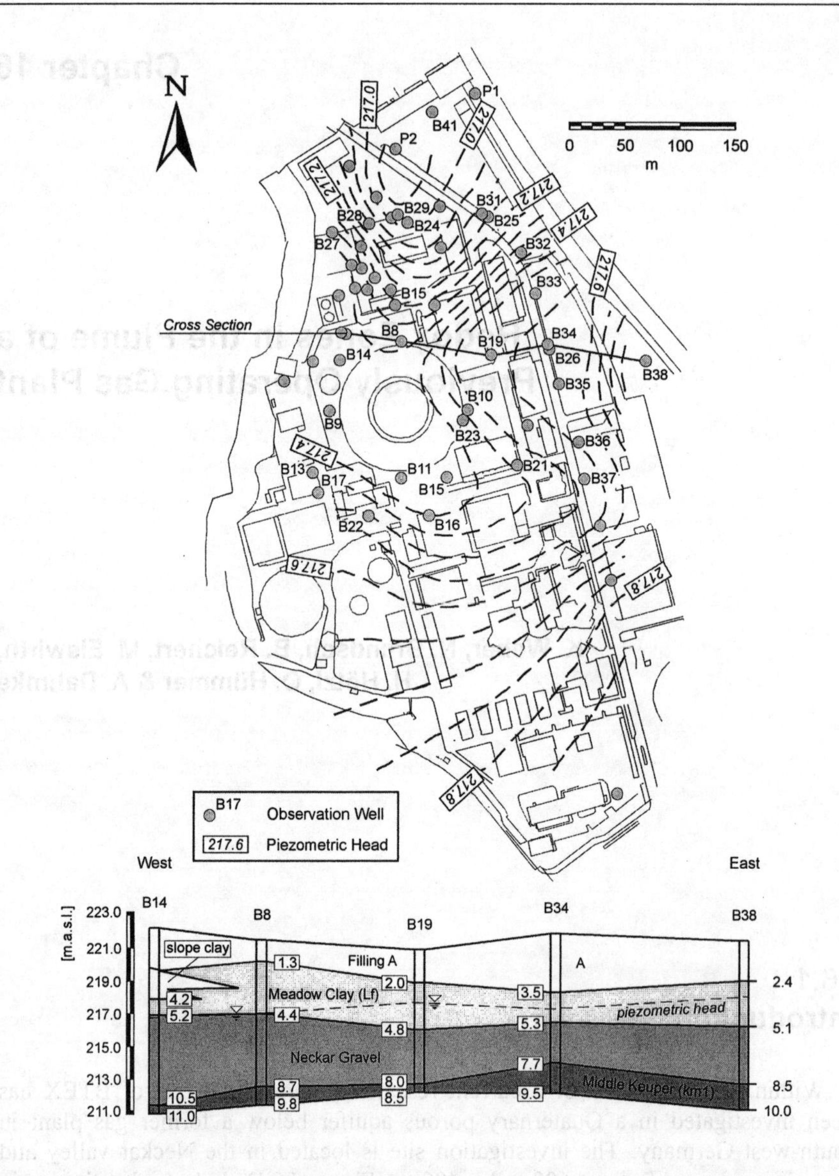

Fig. 16.1: Investigation area with the location of groundwater observation wells; potentiometric surface at the test site in August 1997 and schematic geological cross section (modified from EISWIRTH et al., 1997).

BTEX have relatively high pollution potentials, based on their high aqueous solubilities and estimated toxicity or arcinogenity. Thus, our research has concentrated on the behaviour of BTEX in aquifers. To understand the fate of these contaminants in the groundwater, the identification of redox zones in the plume is assumed to be very important. The determination of an exact redox potential is not necessarily acquired, rather an identification of the dominating redox processes as expressed by the electron acceptor being reduced by the BTEX. The redox status of the aquifer can be assigned on the basis of analyses of free oxygen, nitrate, nitrite, ammonium, manganese, iron, sulfate, and sulfide in groundwater.

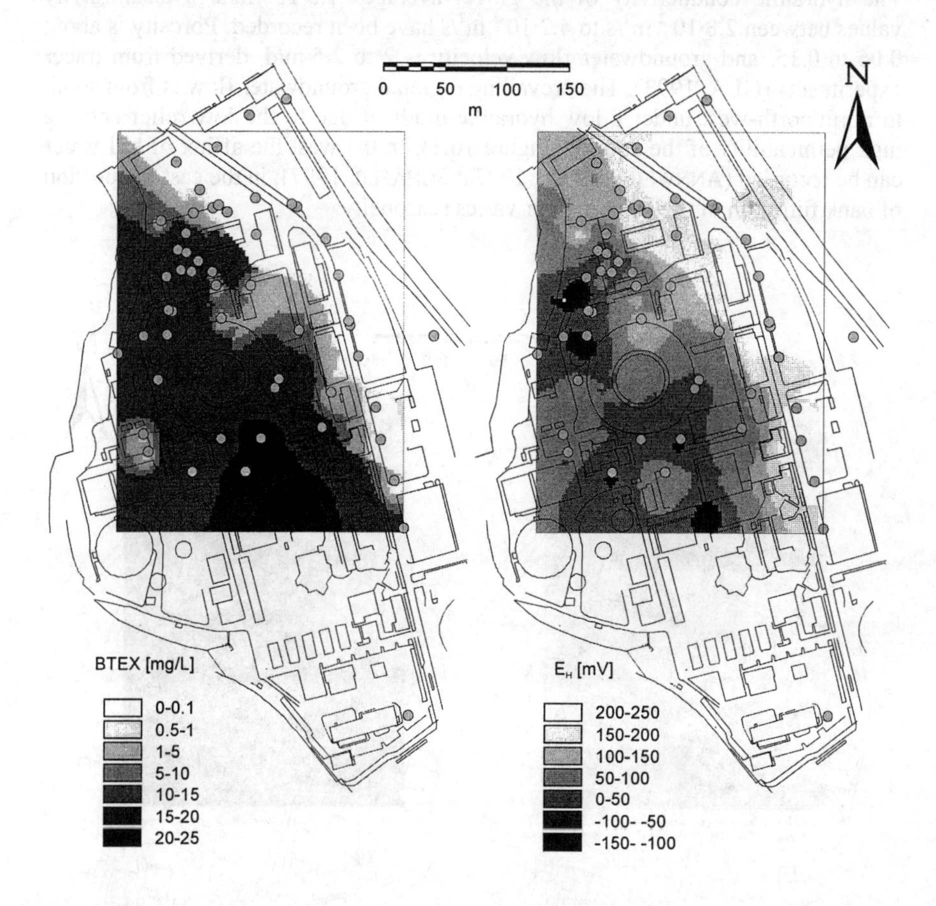

Fig. 16.2: Concentration of BTEX and E_H (SHE) one meter below groundwater level at the investigation site (August 1997).

16.2 Geology and Hydrology

The aquifer at the test site consists of medium-grained gravel of Quaternary age intersected locally by sandy and silty lenses. This porous aquifer is underlain by a sequence of Upper Triassic rocks (Middle Keuper), which dips gently north to north-west at very low degrees. The whole sequence is locally affected by east-west striking faults that have partially disrupted the Triassic rocks. The Quaternary gravel aquifer is about 3.5 m in thickness and is overlain by 1 to 2 m thick flood clays and silts as well as anthropogenic fillings (2 to 4 m in thickness). Groundwater is confined in the north and unconfined in the south of the investigation site. The hydraulic conductivity of the gravel averages $1.5 \cdot 10^{-3}$ m/s. Transmissivity values between $2.8 \cdot 10^{-2}$ m²/s to $4.2 \cdot 10^{-5}$ m²/s have been recorded. Porosity is about 0.06 to 0.15, and groundwater flow velocity is 2 to 2.5 m/d, derived from tracer experiments (GLA, 1993). The prevailing regional groundwater flow is from south to north/north-west under a low hydraulic gradient due to the low relief and the high permeability of the aquifer (Figure 16.1). In the west the afflux of hill water can be recorded (ANNWEILER et al., 1997a; SCHÄFLE, 1997); in the east the portion of bank filtration of the Neckar river varies seasonally.

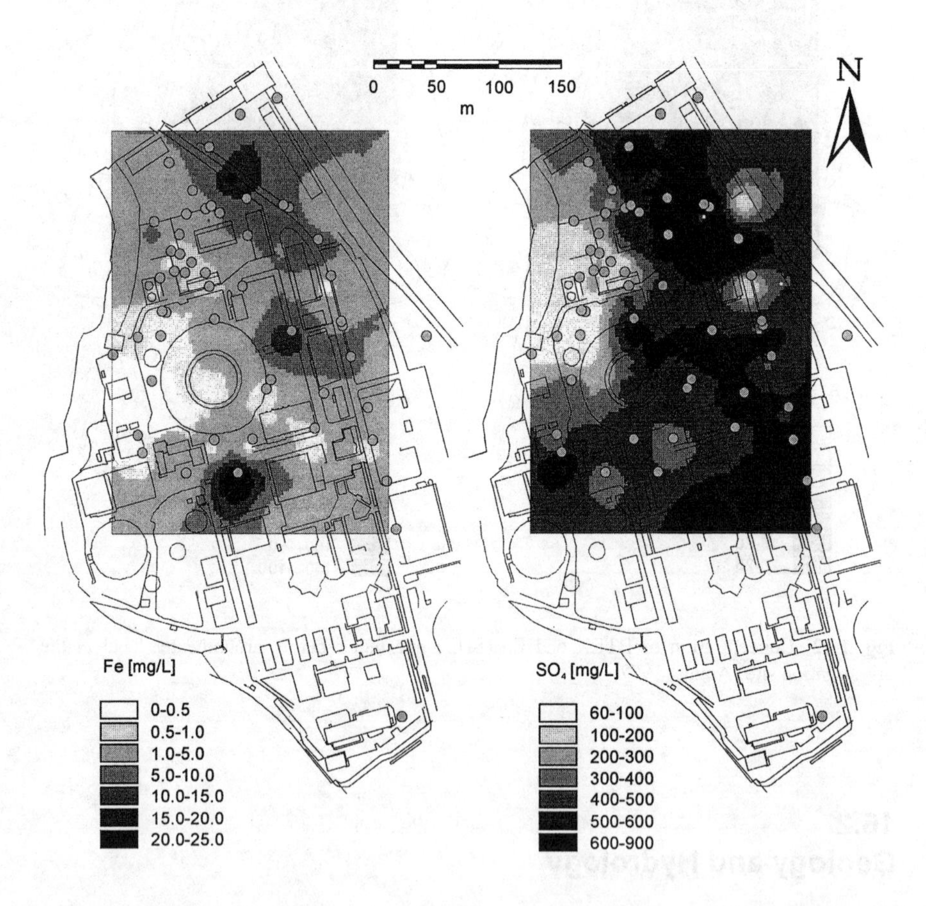

Fig. 16.3: Concentration of dissolved iron and sulfate one meter below groundwater level at the investigation site (August 1997).

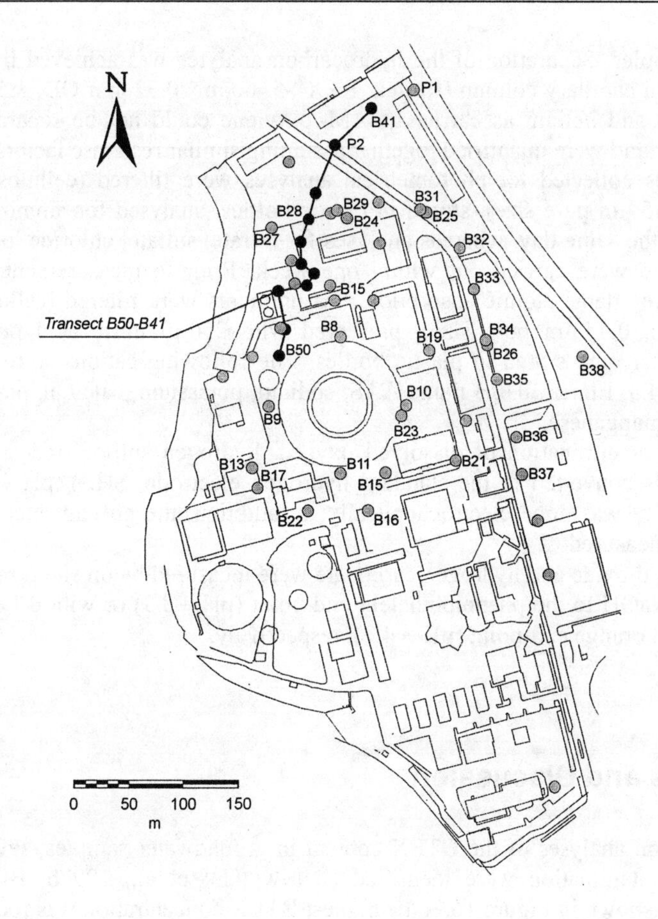

Fig. 16.4: Site map showing Transect B50-B41 (modified from BRANDSCH, 1998).

16.3 Methods and Materials

In order to investigate the relation between the decrease of aromatic hydrocarbons and the increase in the redox potential of the groundwater due to variations in the ferrous iron and sulfate concentrations, 59 groundwater observation wells have been sampled along the aquifer flowpath. Fourteen of these wells have been constructed with multilevel-sampling-systems (HERFORT et al., 1998), allowing the taking of samples in 4 to 8 horizons.

To determine the BTEX concentration in groundwater, 20-ml glass vials were filled with 10 ml of groundwater and capped with teflon-lined septa. Analyses were performed on the same day using a Hewlett-Packard HP5890 gas chromatograph equipped with a flame ionisation detector connected to a HP19395A headspace sampler. Separation of the hydrocarbon analytes was achieved by using a fused silica capillary column (Restek, RTX®-5, 60 m x 0.32 mm OD, 1.5 µm film thickness) and helium as carrier gas. M-/p-xylene could not be separated from each other and were quantified together assuming similar response factors.

Samples collected for photometrical analyses were filtered (cellulose acetate filters, 0.45 µm pore size), stored in plastic bottles, analysed for ammonium and nitrite on the same day, whereas analyses for nitrate, sulfate, chloride, phosphate, and silicate were carried out within one week. Prior to measurement, samples prepared for flame atomic absorption measurements were filtered (cellulose acetate filters, 0.45 µm pore size), preserved with 250 µl nitric acid per 100 ml groundwater, and stored in plastic bottles. The following cations were analysed with a Perkin Elmer 3030B flame AAS: sodium, potassium, calcium, magnesium, iron, and manganese.

In situ determination of dissolved oxygen, hydrogen sulfide, redox potential (afterwards converted to the standard hydrogen electrode, SHE), pH-value, and conductivity was done electrochemically; in addition, the groundwater temperature was measured.

Carbon dioxide and hydrogen carbonate were measured by on site titration with 0.025 N NaOH to the phenolphthalein end-point (pH = 8.3) or with 0.1 N HCl to the methyl orange end-point (pH = 4.2), respectively.

16.4 Results and Discussion

Based on analyses of the BTEX content in groundwater samples, several centres of contamination were identified (ANNWEILER et al., 1997b; BRANDSCH, 1998). As shown in Figure 16.2, the highest BTEX concentration was found in the southern part of the investigation site. Aromatic hydrocarbons were not detected in the well (B58) upgradient from the contamination source. Farther downgradient, up to 25 mg/L of BTEX were analysed in the groundwater zone which was identified to possess the highest reducing capacity. Horizontally, the plume of aromatic hydrocarbons extends about 300 m downgradient from the contamination source, decreasing in concentration rapidly along the aquifer flowpath. This is to be expected from the dilution of the plume, degradation (BARKER et al., 1987; KAO & BORDEN, 1997; EDWARDS et al., 1992) and, if the plume is not in a steady state, from sorption effects. Concentrations of BTEX in the contaminant plume and the lateral extent of the plume have remained stationary at least since summer 1997, indicating that microbial degradation and sorption processes have come into a dynamic equilibrium with the delivery of contaminants.

The E_H-contours are also depicted in Figure 16.2. Redox potentials varied over a range of +250 to –100 mV and appeared to be influenced by the degree of hydrocarbon contamination. Based on these redox potentials, it appears that site specific redox conditions range from moderate to highly reducing.

As shown in Figure 16.3, redox potential contours match quite well with the concentration distribution of dissolved iron in groundwater. At the prevailing redox condition in the high contaminated groundwater, Fe(III) oxides, which are

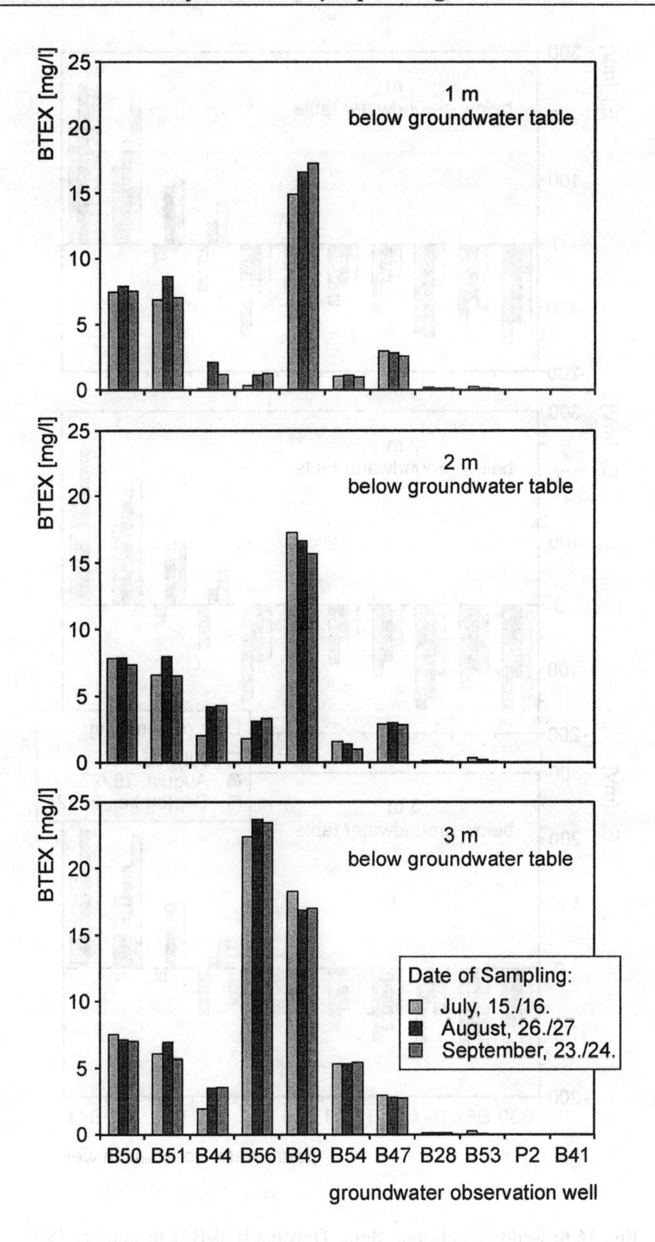

Fig. 16.5: BTEX concentrations along Transect B50-B41 in summer 1997.

abundantly present but usually insoluble in aquifers, were reduced to ferrous iron. In addition, sulfate reduction is indicated by significant sulfide content of the contaminated groundwater and a decrease in sulfate concentrations in the north-western part of the investigation site. The content of dissolved iron and sulfate in

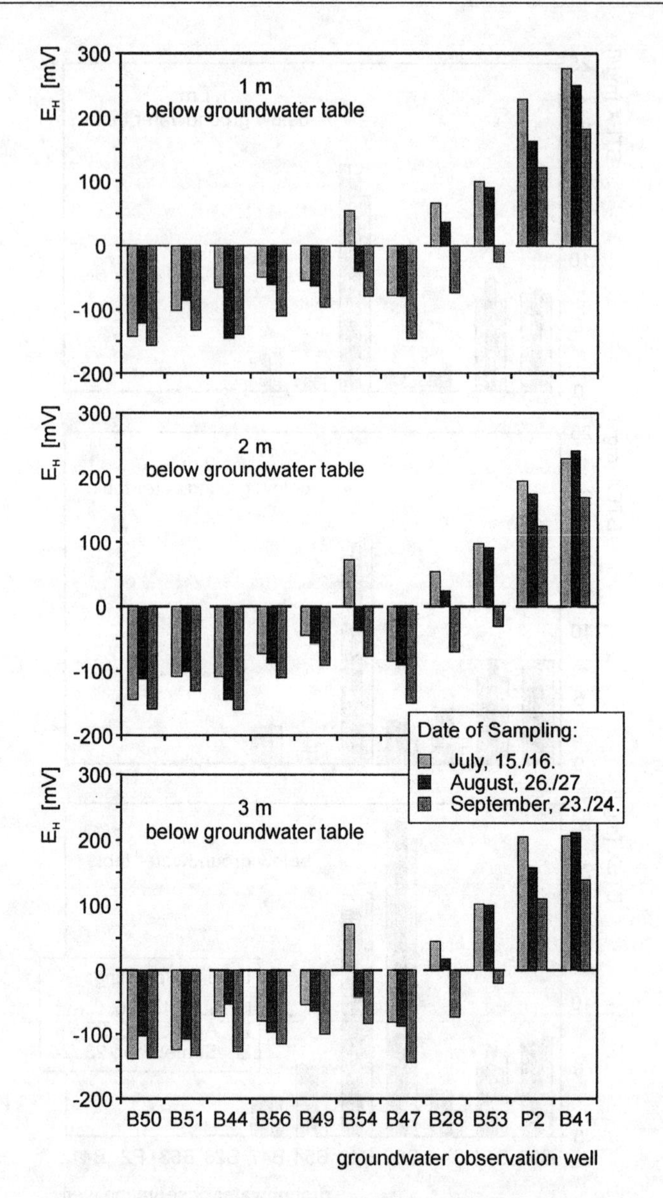

Fig. 16.6: Redox conditions along Transect B50-B41 in summer 1997.

the north-western part of the site indicates that reduced sulfur species probably react with the reduced iron, precipitating as iron sulfides. Therefore, the concentrations of dissolved iron and sulfate increase simultaneously along the aquifer flowpath. Besides iron and sulfate reduction, nitrate reduction is indicated by significant decrease in concentration in the centre of the contaminant plume.

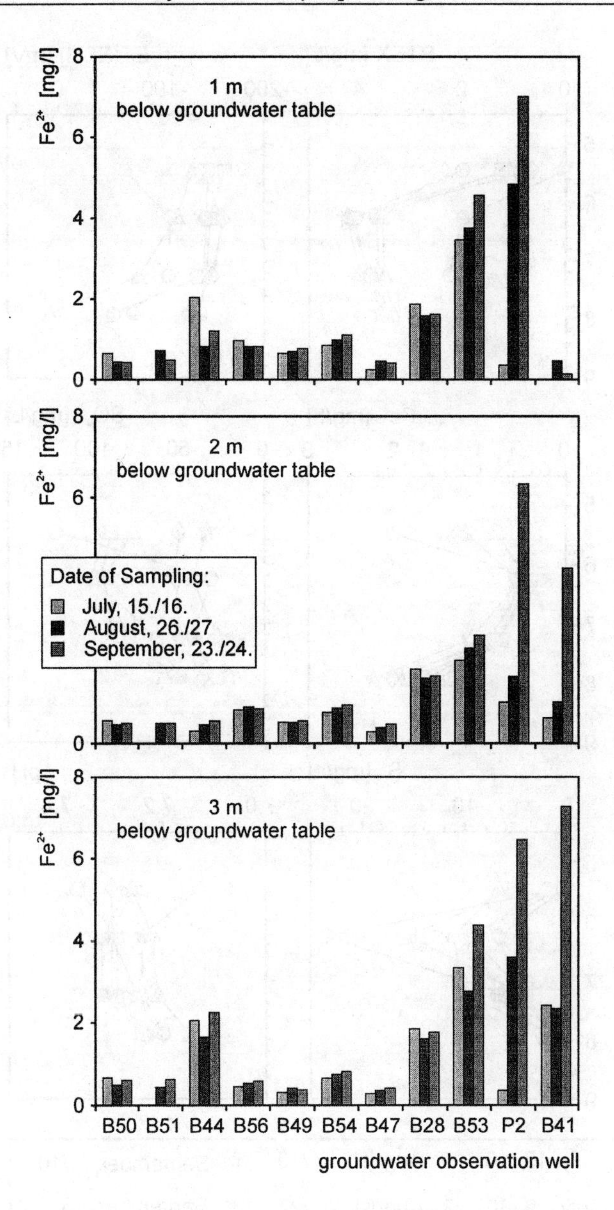

Fig. 16.7: Concentrations of dissolved iron along Transect B50-B41 in summer 1997 (modified from BRANDSCH, 1998).

Nitrate reduction can only play a minor role in the biodegradation process, because the nitrate concentration is only about 24 mg/L upgradient from the contamination source. Due to the high sulfate concentrations in the groundwater, sulfate reduction is the predominant terminal electron-accepting process. One reason for those high sulfate concentrations is the leaching of the anthropogenic

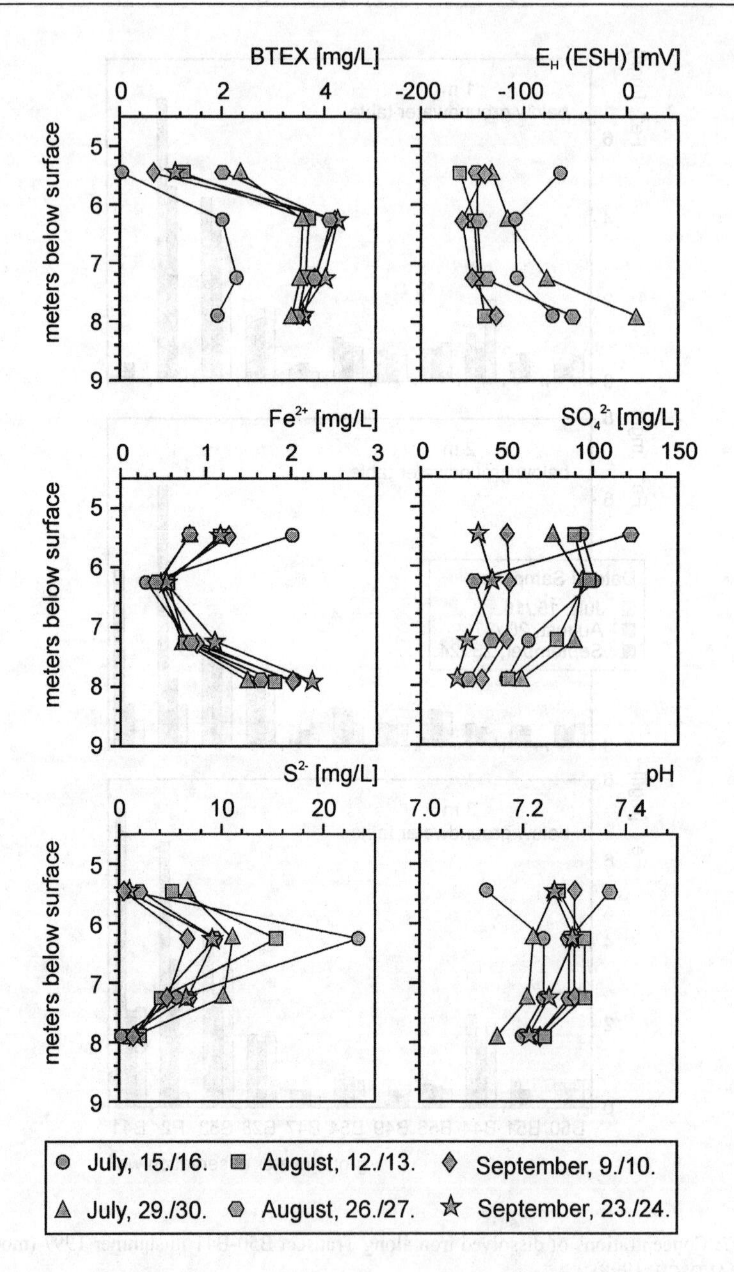

Fig. 16.8: Geochemical conditions at observation well B44.

fillings (2 to 4 m in thickness) presented nearly everywhere at the test site. Besides that the ascent of highly mineralised water along an east-west striking fault in the north-eastern part of the investigation site delivers unusually large amounts of sulfate to the aquifer system. For an explanation of the enhanced iron concentrations in the highly contaminated groundwater in the south of the site, two different hypotheses are discussed (SCHMITT et al., 1996). Based on the laboratory study of BELLER et al. (1992) the enhanced iron concentrations can be a secondary, presumably abiotic, product of the biodegradation of aromatic hydrocarbons coupled to sulfate reduction. After LOVELY et al. (1994) an increased bio-availability of the ferric compounds might be the result of a complexation between organic acids, the metabolites of BTEX, and insoluble Fe(III) oxides.

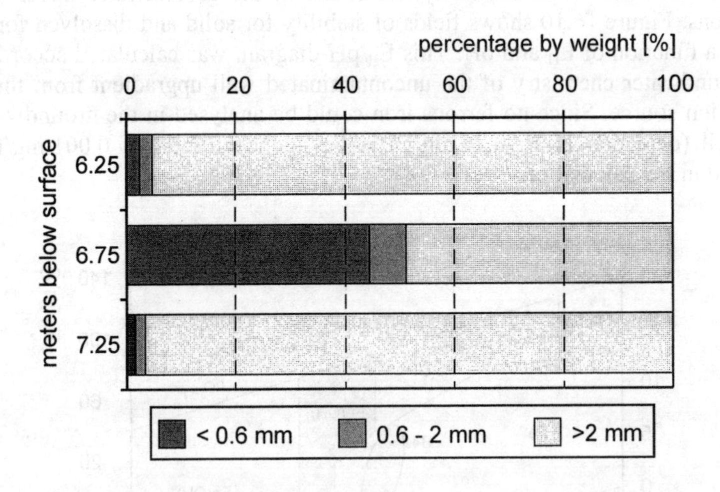

Fig. 16.9: Grain size analysis of sediments at observation well B44.

In order to obtain more specific information as to the stability of the pollutant compounds underground, groundwater samples were collected in 4 to 8 horizons along the transect shown in Figure 16.4. The BTEX concentrations vary vertically as well as laterally, while no BTEX were found in the two observation wells most downgradient from the contamination centre (Figure 16.5). As expected and mentioned above, the E_H of the groundwater increases with decreasing BTEX concentrations (Figure 16.6). A seasonal variation of the redox potential can be seen in Figure 16.6. The drop of the redox potential can be attributed to an increase of dissolved iron (Figure 16.7) during the observation period. The constant low iron concentrations in the wells B50 to B47 indicate that sulfur reduction predominates in this groundwater region and therefore most of the reduced iron precipitates as iron sulfides. More downgradient, the reduced iron concentrations increase to the upper part of the groundwater in B41, because, owing to the prevailing redox condition, no Fe(III) oxides were reduced any further.

A possible reason for a higher redox potential in the upper part of B41 than further downward may be that dissolved oxygen is carried to contaminated sediments by recharge from atmospheric precipitation, but is quickly utilised by aerobic degradation processes. Generally, vertical changes in the geochemistry of the groundwater could be correlated with anomalies in the geological structure of the aquifer. This fact can be demonstrated by the geochemistry of the observation well B44 (Figure 16.8 and Figure 16.9). At 6.75 m below ground level, the percentage of finest grain sizes is higher than in the other horizons of B44. Due to sorption effects, the BTEX concentration in the groundwater of the 6.75 m level is the highest. The highest sulfide concentrations were measured in the most contaminated area, whereas redox potential and the concentration of dissolved iron were at the lowest level. The correlation of organic and inorganic parameters indicates biodegradation of the aromatic hydrocarbons in the groundwater under anoxic conditions. Figure 16.10 shows fields of stability for solid and dissolved forms of iron as a function of E_H and pH. This E_H/pH-diagram was calculated according to the groundwater chemistry of the uncontaminated well upgradient from the contamination source. Since no ferrous iron could be analysed in the groundwater of that well (detection limit: 0.25 mg/L), a Fe^{2+}-concentration of 0.001 mg/L was assumed in the calculation.

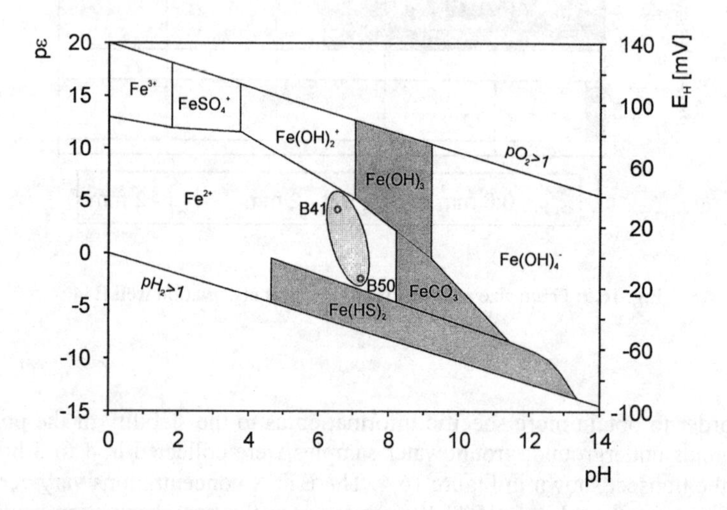

Fig. 16.10: Stability fields of solid and dissolved iron species as a function of E_H and pH at 21°C. The E_H/pH-diagram was calculated according to the groundwater chemistry of the uncontaminated well up-gradient from the contamination source. The water composition is: Fe = 0.02 μM, Ca = 10.13 mM, Mg = 2.47 mM, Na = 3.48 mM, K = 0.35 mM, Mn = 23.8 μM, Si = 78.4 μM, Cl = 5.81 mM, S = 8.93 mM, N = 0.43 mM, P = 0.1 μM and IC = 8.76 mM.

Stability regions for the solids are shaded and the predominant dissolved forms are indicated in the various unshaded areas. The sloping upper and lower boundaries within the diagram are limits within which water itself is chemically stable. Above the upper boundary, water is oxidised to yield oxygen gas and is, below the lower boundary, reduced to yield hydrogen gas. The circled area in Figure 16.10

indicates the position of the plume-derived groundwater within the E_H/pH-diagram. Exemplary for the hydrochemical situation at the contamination centre and leading beyond the actual contamination zone, the E_H/pH-values of the groundwater at B50 and B41 at 5 m below ground level are depicted in the diagram.

As Figure 16.10 demonstrates, the pH value of the groundwater increases with decreasing E_H. This observation can be explained with an increased reduction of iron(III) hydroxides resulting in a higher amount of hydroxide ions (GIBSON et al., 1998). Besides, the iron species of the groundwater, dominating in the E_H/pH-diagram, match quite well with the iron concentrations analysed in the groundwater of the corresponding observation wells. Consequently, the E_H/pH-value gives a first impression of the dominating redox-sensitive processes in groundwater systems.

16.5 Summary and Conclusions

Leakage of BTEX from a former gas plant into a mineralised and slightly oxidising aquifer environment causes a drastic change in the inorganic aqueous chemistry of the contaminant plume.

A sequence of redox zones is identified on the basis of groundwater chemical analyses. In the centre of contamination, sulfidogenic conditions prevail, followed by ferrogenic, nitrate-reducing, and aerobic environments over a distance of 300 m. The redox zone sequence is consistent with thermodynamical principles and closely matches with the BTEX plume.

The redox potential of groundwater is a useful tool to obtain a first impression of the dominant redox-sensitive processes in the aquifer system. Since the electrochemical determination of redox potentials is mechanically very simple, this parameter can be used in the long-term supervision of a contaminated test site and for monitoring purposes applied in addition to other hydrochemical measurements.

16.6 References

ANNWEILER, E.; EISWIRTH, M.; HÖTZL, H.; REICHERT, B.; RICHNOW, H.H.; SCHÄFLE, B.; SEIFERT, R. & MICHAELIS, W. (1997A): Organische Kontaminationen im Grundwasserbereich eines ehemaligen Gaswerkstandortes (Organic contamination in an aquifer aerea of a former gas plant). In: KREYSA, G. & WIESSNER, J. (Eds.) Möglichkeiten und Grenzen der Reinigung kontaminierter Grundwässer – Resumee und Beiträge des 12. Dechema-Fachgesprächs Umweltschutz, Brönners Druckerei Breidenstein GmbH, Frankfurt, 549-556.

ANNWEILER, E.; EISWIRTH, M.; HÖTZL, H.; MICHAELIS, W.; PAPE, T.; REICHERT, B. & RICHNOW, H.H. (1997b): Distribution, transport and biodegradation of organic contaminants in a near surface aquifer in the area of a manufactured gas plant site. Proceedings of the EUG-Conference, March, Straßburg, 23-27.

BRANDSCH, N. (1998): Milieuparameter im Grundwasser in einer BTEX-Abstromfahne (Milieu indicators in groundwater in a BTEX plume). unpublished Diplom thesis, University of Karlsruhe, 72 p.

BARKER, J.F.; PATRICK, G.C. & MAJOR, D. (1987): Natural attenuation of aromatic hydrocarbons in a shallow sand aquifer. Groundwater Monitoring Review 7/1: 64-71.

BELLER, H.R.; GRBIC-GALIC, D. & REINHARD, M. (1992): Microbial degradation of toluene under sulfate reducing conditions and the influence of iron on the process. Appl. Environm. Microbiol. 58: 786-793.

EDWARDS, E.A.; WILLS, L.E.; REINHARD, M. & GRBIC-GALIC, D. (1992): Anaerobic degradation of toluene and xylene under sulfate reducing conditions. Appl. Environm. Microbiol. 58: 794-800.

EISWIRTH, M.; HÖTZL, H.; REICHERT, B. & WEBER, K. (1997): Field soil gas screening methods for delineation of subsurface contamination. In: GOTTLIEB, J.; HÖTZL, H.; HUCK, K. & NIESSNER, R. (Eds.) Field screening Europe. Kluwer Academic Publishers, Dordrecht Bosten London, pp. 29-32.

GIBSON, T.L.; ABDUL, A.S. & CHALMER, P.D. (1998): Enhancement of in situ bioremediation of BTEX-contaminated ground water by oxygen diffusion from silicone tubing. Groundwater Monitoring Review 18/1: 93-104.

GLA (Geologisches Landesamt, Baden-Würtemberg) (1993): Unpublished hydrogeological report of the Geological Survey, Az.: 2565.02/93-4764 Ra/Ro7Szi, 29.10.1993, Stuttgart.

HERFORT, M.; PTAK, TH.; TEUTSCH, G.; HÜMMER, O. & DAHMKE, A. (1998): Testfeld Süd: Einrichtung der Testfeldinfrastruktur und Erkundung hydraulisch-hydrogeochemischer Aquiferparameter (Test site South: Infrastructural development and determination of hydraulic and hydrogeochemical aquifer parameters). Grundwasser 3/3, (in press).

HESKE, C. (1998): Erfassung kurzzeitiger Schwankungen im Bodenlufthaushalt über BTEX-Abstromfahne (Determination of short term variations in the soil gas zone above a BTEX plume). unpublished Diplom thesis, University of Karlsruhe, 74 p.

LOVELY, D.R.; WOODWARD, J.C. & CHAPELLE, F.H. (1994): Stimulated anoxic biodegradation of aromatic hydrocarbons using Fe(III) ligands. Nature 370: 128-131.

KAO, C.-M. & BORDEN, R. (1997): Site-Specific Variability in BTEX Biodegradation Under Denitrifying Conditions. Groundwater 35/2: 305-311.

POSTMA, D. & JAKOBSEN, R. (1996): Redox zonation: Equilibrium constraints on the Fe(III)/SO4-reduction interface. Geochim. Cosmochim. Acta 60/17: 3169-3175.

SCHÄFLE, B. (1997): Regionale und saisonale Veränderungen im Grundwasser der TWS (Regional and seasonal variations in groundwater in the test site TWS). unpublished Diplom thesis, University of Karlsruhe, 59 p.

SCHMITT, R.; LANGGUTH, H.-R.; PÜTTMANN, W.; ROHNS, H. P.; ECKERT, P. & SCHUBERT, J. (1996): Biodegradation of aromatic hydrocarbons under anoxic conditions in a shallow sand and gravel aquifer of the Lower Rhine Valley, Germany, Org. Geochem. 25: 41-50.

Chapter 17

Degradation of Organic Groundwater Contaminants: Redox Processes and E_H-Values

M. Ebert, O. Hümmer, M. Mayer, O. Schlicker & A. Dahmke

17.1 Introduction

In many cases the E_H-value of a groundwater sample is either measured or calculated from the distribution of certain redox couples, with the result that the utility of the resulting value is small. The selective sensitivity of the redox electrodes (LINDBERG & RUNNELLS, 1984), partial non-equilibrium and the resulting mixture of potentials (STUMM & MORGAN, 1996), or an insufficient preparation of the probes (e.g. KÖLLING, 1986) are made responsible for the uncertainty of this "weak" milieu parameter. Especially in sulfidic environments often too high E_H-values are found. The insensitivity of the platinum redox electrodes for the sulfide/sulfate redox couple are surely responsible for this.

The main purpose for determining the redox potential in groundwater samples is to deduce the dominant redox milieu in the subsurface and to forecast possible redox reactions. In a narrower sense, the E_H-value is used to confine the probable processes and reaction mechanisms. A more exact method to deduce the dominant redox process in an aquifer is the sequential extraction of the solid phase (e.g. HERON, 1994a) to determine the distribution of the redox species in the matrix. In most cases, however, only water samples are available which integrate over a larger area of the aquifer. Thus, it is necessary to deduce the relevant redox mechanisms from the analysis of groundwater samples, as it has been done several times for laboratory column experiments (EBERT, 1997; HENCKE, 1998; VON GUNTEN & ZOBRIST, 1993).

Redox reactions may play a major role by the degradation of organic contaminants in aquifers. In principle, contaminants which are in a reduced state can be distinguished from such contaminants which are oxidised. In Figure 17.1, the standard potential of the degradation steps of some common groundwater contaminants are compared with the standard potentials of some inorganic redox couples, which are important in groundwater milieus. It is easy to recognise that the contaminants of the BTX-group (benzene, toluene, xylene) can be broken down to carbon dioxide through oxidation processes, whereas the halogenated aliphatic hydrocarbons act as strong oxidising agents, which can be dehalogenated in several reductive reaction steps. Depending on the differences in the potentials, several possible reaction partners exist in groundwater environments for both reaction mechanisms. Here, the examples BTX- and VOC-degradation (BTX: Benzene, T: Toluene, X: Xylene; VOC: volatile organic carbons) shall be used to show the interaction between the redox potential in the pore water and the actual degradation reaction, to verify the usefulness of E_H-measurements in these special environments.

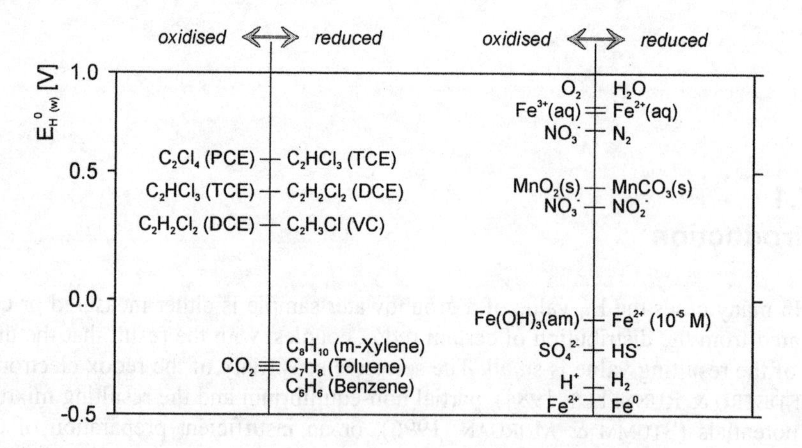

Fig. 17.1: Standard potential in water at pH = 7 ($E_H{}^0{}_{(w)}$) of degradation steps of common organic groundwater contaminants and transformation of inorganic redox species (modified from DAHMKE et al., 1997).

17.2 Degradation of BTX in the Underground of a Former Gas Plant

The degradation of benzene, as an example for BTX-degradation, is shown in Equations (17.1) to (17.5).

$$\begin{aligned} 7.5O_2 + C_6H_6 &\Rightarrow 6CO_2 + 3H_2O \\ 6NO_3^- + C_6H_6 + 6H^+ &\Rightarrow 6CO_2 + 6H_2O + 3N_2 \\ 15MnO_2 + C_6H_6 + 30H^+ &\Rightarrow 6CO_2 + 18H_2O + 15Mn^{2+} \\ 30Fe(OH)_3 + C_6H_6 + 60H^+ &\Rightarrow 6CO_2 + 78H_2O + 30Fe^{2+} \\ 3.75SO_4^{2-} + C_6H_6 + 7.5H^+ &\Rightarrow 6CO_2 + 3H_2O + 3.75H_2S \end{aligned} \quad (17.1\text{-}5)$$

The degradation process is catalysed by microbes. The literature describes different micro-organisms which are capable of degrading BTX with all common electron acceptors (i.e. ZEYER et al., 1986; LOVLEY & LONERGAN, 1990; EDWARDS et al., 1992; EDWARDS & GRIBIC-GALIC, 1994). Aerobic conditions are the best requirements for fast BTX degradation similar to the microbial catalysed oxidation of other organic substances. In the anaerobic environment the Fe(III) solid phases are considered as the most important electron acceptors and, besides, the natural oxidation capacity of an aquifer is due to the Fe(III) content (HERON et al., 1994a).

The test site which is discussed here is a former gas plant in the south of Germany with a BTX contamination in the subsurface (see also Chapter 16). With increasing distance downstream from the main contamination source there are lower BTX concentrations in the groundwater. This indicates, together with the concentration profiles of other groundwater constituents (namely iron, secondary sulfate and sulfide) as well as with the pH- and E_H-values, the existence of microbial degradation processes. The E_H-values in groundwater indicate that the degradation is mainly governed by the reduction of Fe(III) solids. Figure 17.2 shows a pε/pH-diagram in which the pε-values of the test site groundwater samples are plotted. The basis of the diagram calculation with PHREEQE (PARKHURST et al., 1990) is an average groundwater. The curves represent a saturation index of zero for different iron minerals, as suggested by EBERT (1997). The plotted pε/pH-values of the groundwaters mainly indicate a dominant iron redox system. The majority of the solutions are nearly equilibrated with amorphous Fe(III)hydroxide and siderite, what is typical for environments with microbial degradation of organic substances by means of Fe(III) reduction.

Conversely, the results of chemical extractions of the aquifer material show in some cases the influence of a microbial sulfate reduction. The used extraction methods (modified from HERON et al., 1994a) by employing several aggressive extractants resulted not only in total Fe(II) and Fe(III) concentrations, but the iron content can be allocated to the nominal binding forms of low crystalline and crystalline Fe(III), monosulfidic, pyritic and carbonatic/phosphatic bounded Fe(II).

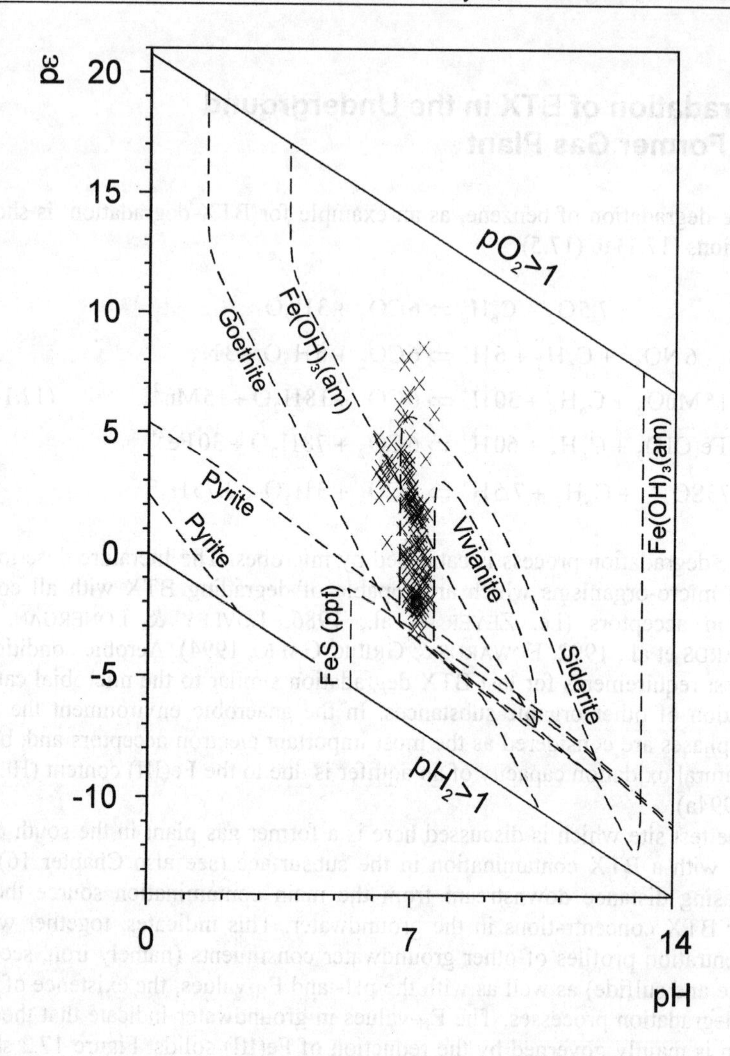

Fig. 17.2: pε/pH-diagram (iron solids) for an intermediate composition of the test site groundwater at 15°C and 1 bar. The curves show the saturation index SI = 0 and were calculated with PHREEQE (PARKHURST et al., 1990). The water composition is [mmol/L]: Na = 7.0, K = 0.5, Mg = 2.5, Ca = 7.0, Fe = 0.05, Mn = 0.01, Cl = 8.0, S = 5.0, C = 8.0, N = 0.35, Si = 0.1 and P = 0.005.
Methane is not included in the data set. The crosses represent the pε/pH-values of all groundwater samples taken from December 1996 to June 1997.

Moreover, the sulfur content may be divided in the formal binding forms acid volatile sulfur (AVS, mainly FeS-sulfur) pyrite sulfur and the total sulfur content without the pyrite fraction.

Figure 17.3 shows the total Fe(II) content and the concentrations of the different sulfur binding forms of aquifer samples (fraction <2 mm) of representative locations along a transect in groundwater flow direction. It is shown that, downstream from the contamination source, the main part of Fe(II) is bound in sulfidic forms. Considering that just one electron was necessary for the reduction of each bounded Fe(II) and that for each sulfidic bounded sulfur 14 (FeS_2) or 8 (FeS) electrons have to be transferred, it is apparent that sulfate reduction is potentially an important factor of the BTX-degradation at the test site. However, these facts are not supported only by the measurement of the redox environment of the groundwater samples.

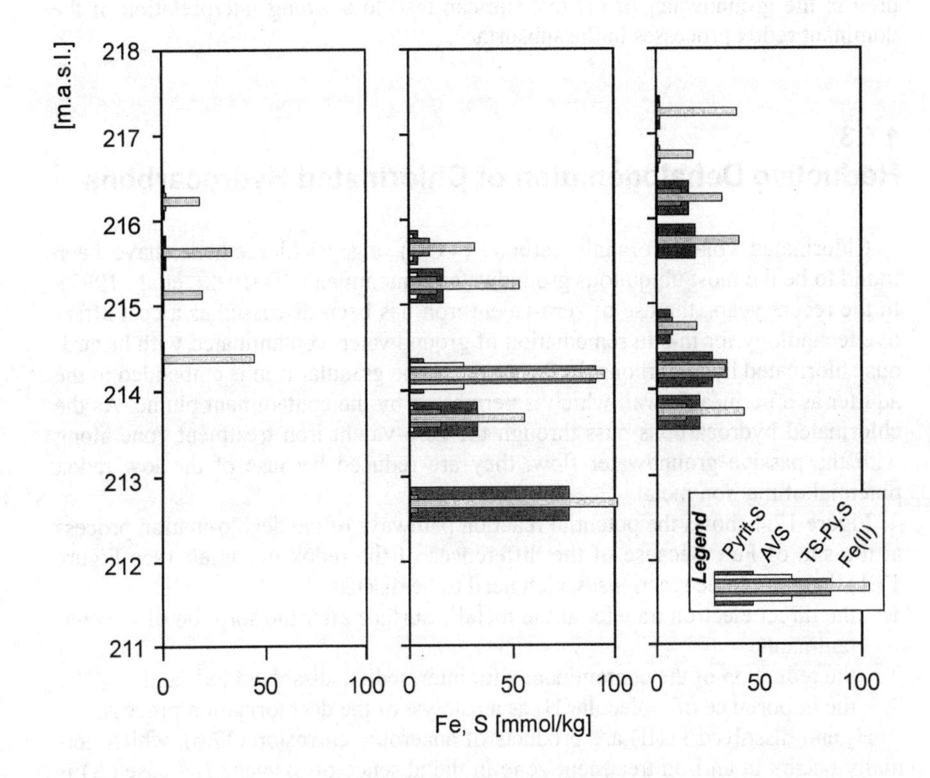

Fig. 17.3: Total Fe(II)- and S-contents in aquifer samples (fraction <2 mm) from the test site at the upstream location (well B58, see Chapter 16) and downstream the main contamination zone (B50, B47). The sulfur species are pyrite bounded (Pyrite-S), acid volatile sulfur (AVS: monosulfides) and total sulfur without the pyrite content (TS-Py.S).

The difference between the redox environments deduced from the solid phase extractions or the groundwater E_H-values cannot be explained only by the lower sensitivity of the redox electrodes. Some recent groundwater data show only lower sulfide concentrations (<12 mg/L). Moreover, at some local points, sulfide could not be detected in the groundwater even if sulfidic bounded sulfur was measured in the solid phase at these sections of the aquifer. A parallel occurrence of Fe(III) and sulfur reduction in the aquifer may explain the different redox signals if the fast precipitation kinetic of amorphous iron sulfides are taken into account. With a surplus of dissolved Fe(II) in the groundwater, sulfides will precipitate quickly so that sulfide would not accumulate in the solution. Therefore, the products of the microbial sulfate reduction could not affect the E_H to a higher degree and the Fe(II) content of the solution remains responsible for the measured E_H-value. In this case, the extent of sulfate reduction can only be proven by the examination of the solid phases. There are similar relationships in aquifers with micro-scale environments where sulfate is reduced. It is concluded that only the E_H-values measured in the groundwater of the test site can lead to a wrong interpretation of the dominant redox processes in the subsurface.

17.3 Reductive Dehalogenation of Chlorinated Hydrocarbons

Chlorinated volatile organic carbons (VOC), e.g. trichloroethene, have been found to be the most ubiquitous groundwater contaminants (DAHMKE et al., 1997). In the recent years, the use of zero-valent iron has been discussed as a cost effective technology for in-situ remediation of groundwater, contaminated with hazardous chlorinated hydrocarbons. In this concept, the granular iron is embedded in the aquifer as a permeable wall which is permeated by the contaminant plume. As the chlorinated hydrocarbons pass through the zero-valent iron treatment zone along with the passive groundwater flow, they are reduced because of the low redox potential of the iron metal.

Figure 17.4 shows the potential reaction pathways of the dechlorination process at the site of Fe^0. Because of the differences of the redox potentials (see Figure 17.1) there are three reactions, which need to be discussed:

1. the direct electron transfer at the metal's surface after the sorption of the contaminants,
2. the reduction of the contaminant with intermediate dissolved Fe^{2+} and
3. the importance of molecular H_2 as a catalyst of the dechlorination process.

H_2 and dissolved Fe(II) are products of anaerobic corrosion (17.6), which normally occurs in an iron treatment zone in the absence of oxygen. The case (A) is the quantitatively important reaction pathway for the dehalogenation of VOC. The dehalogenation by dissolved Fe(II) or by H_2 is only secondarily significant, because the reaction kinetics are comparatively slow (SWEENY & FISCHER, 1972; MATHESON & TRATNYEK, 1994; GILLHAM & O'HANNESIN, 1994).

The stepwise dechlorination of trichloroethene (TCE) in the presence of Fe^0 according to (17.7), (17.8) and (17.9), occurs in the intermediate build-up of less chlorinated ethenes like dichloroethene (DCE), (17.7), and monochloroethene (VC), (17.8). However, hardly any release of these chlorinated intermediates could be detected, only the final product ethene was observed in a higher extent (17.9).

$$2H_2O + 2Fe^0 \Leftrightarrow H_2 + 2Fe^{2+} + 2OH^- \qquad (17.6)$$

$$C_2HCl_3 + H^+ + Fe^0 \Leftrightarrow C_2H_2Cl_2 + Cl^- + Fe^{2+} \quad (17.7)$$

$$C_2H_2Cl_2 + H^+ + Fe^0 \Leftrightarrow C_2H_3Cl + Cl^- + Fe^{2+} \quad (17.8)$$

$$C_2H_3Cl + H^+ + Fe^0 \Leftrightarrow C_2H_4 + Cl^- + Fe^{2+} \quad (17.9)$$

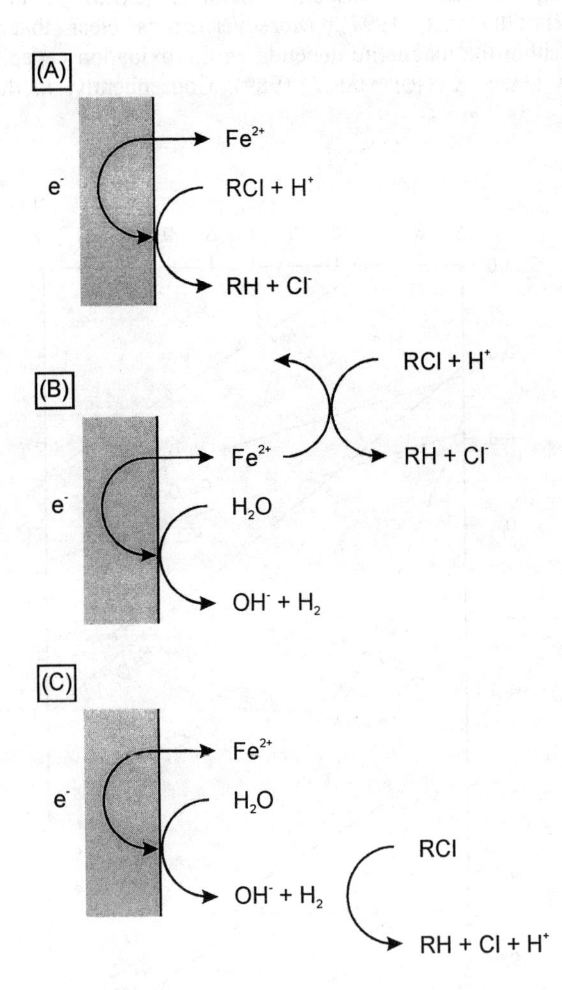

Fig. 17.4: Possible reaction mechanisms of the reductive dechlorination of VOC in the Fe^0/H_2O system. The reaction (A) is the dominant pathway because the kinetics of the reactions (B) and (C) are too slow.

The measurements of the E_H within the bulk solution of Fe^0 treatment zones (lab or field) detected values between –100 mV and –400 mV and corresponding pH-values between 8 and 11, which agree well with previous studies (SCHUHMACHER, 1995). This E_H/pH-field is shown in the diagram (Figure 17.5) of the system Fe^0/TCE/H_2O (SCHLICKER et al., 1998). These redox conditions indicate the contact of the solutions to the iron oxide magnetite, however, the equilibrium with the iron metal is less probable. These findings agree well with results of corrosion science which detected the predominant occurrence of the cubic spinell magnetite using various spectroscopic methods (LONG et al., 1983; ODZIEMKOWSKI & GILLHAM, 1997). Moreover, it is clear that the ratio of Fe(II)/Fe(III) within the magnetite depends on the oxidation capacity of the bulk solution (STRATMANN & HOFFMANN, 1989). Consequently, in these corrosion

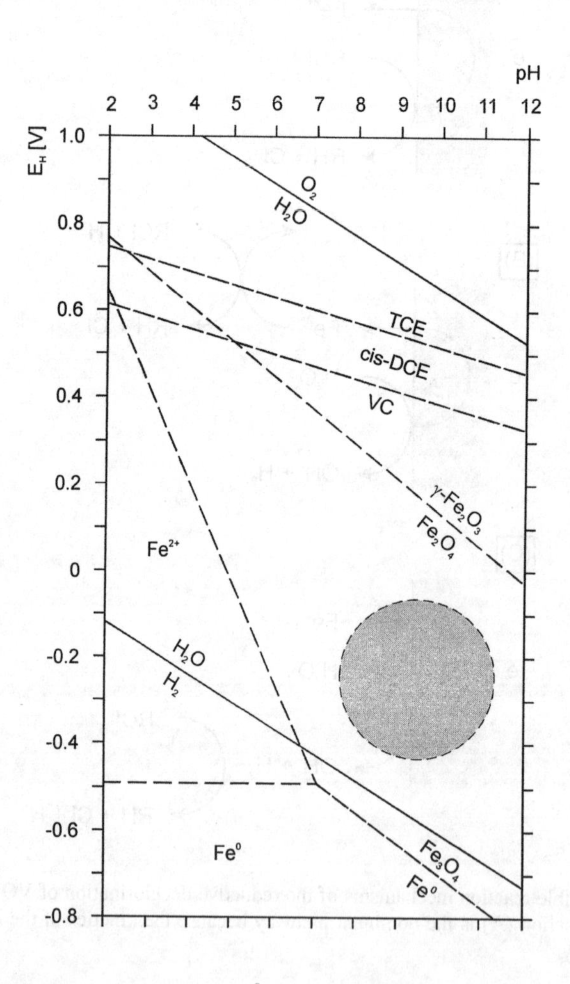

Fig. 17.5: E_H/pH-diagram of the system Fe^0-trichloroethene-H_2O at standard conditions and $[Fe^{2+}] = [Cl^-] = 10^{-3}$ M. The shaded area represents redox conditions typically measured in Fe^0 treatment zones (modified from SCHLICKER et al., 1998).

systems the measured E_H-values seem to be suitable to indicate the real interactions between the solid and the solution (in this case: equilibrium with magnetite). But, it is not magnetite that acts as the final electron donor in the dechlorination of VOC, rather, it is the zero-valent iron metal covered by the magnetite layers.

Magnetite has the lowest electrical resistivity of all iron oxides (CORNELL & SCHWERTMANN, 1996) and electron conduction occurs by unimpeded oscillation between the octahedral Fe^{2+} and Fe^{3+} lattice states (SHUEY, 1975). This high electrical conductivity and the accelerated electron transfer from the iron metal to the boundary (oxide/solution) surface results in a lower redox potential than that of the pure magnetite. Therefore, the reduction of the chlorinated VOC occurs at the magnetite surface with the iron metal as final electron donor. On the other hand, electron transfer is inhibited in the case of precipitation of electrical insulators like vivianite or siderite. This results in a "passivation" of the iron surfaces and therefore in a inhibition of redox processes (HELLMOND & USAKOW, 1992). With a detailed view at the redox system of iron treatment zones, the E_H-values show the interactions between the solution and the solid phase magnetite correctly, but overall, they give the wrong indications to the mechanism of the dehalogenation of VOC at the surface.

17.4 Conclusions

Upon interpreting E_H-values with the aim to characterise the geochemical processes in the subsurface there are a lot of sources for misjudgement. Even if ideal redox electrodes and trouble-free measurements are assumed, the redox potential of the solution cannot always indicate the dominant processes. This is because the E_H cannot be measured at the reaction site, if the reaction occurs on the surface of matrix particles.

In the case of the microbial degradation of BTX in an aquifer, i.e. at the test site, a parallel Fe(III)- and sulfate-reduction may happen without higher sulfide contents in the groundwater, and therefore lower E_H-values are not detectable. If dissolved Fe(II) is present in the system, microbial produced sulfide will precipitate in form of Fe-monosulfides or pyrite. Dissolved sulfide cannot affect the redox potential significantly, because the kinetics of the precipitation are too fast and thus sulfide cannot accumulate in the groundwater. In this case, the amount, or maybe just the existence of microbial sulfate reduction, can only be determined by the detailed analysis of the solid phases.

The E_H/pH-conditions in iron treatment zones correctly indicate an equilibrium with magnetite, but the actual degradation mechanism of chlorinated volatile organic carbons cannot be deduced. At the surface of granular iron covered with magnetite the redox potential of zero valent iron is effective, because of the electrical conductivity of magnetite. Thus, the degradation occurs with iron as the final electron donor while the solution indicates an equilibrium only with regard to the magnetite layers.

17.5 References

CORNELL, R.M. & SCHWERTMANN, U. (1996): The iron oxides. Weinheim VCH-Verlag.

DAHMKE, A.; SCHLICKER, O. & WÜST, W. (1997): Literaturstudie, Reaktive Wände – pH – Redoxreaktive Wände. Berichte LfU.

EBERT, M. (1997): Der Einfluß des Redoxmilieus aus die Mobilität von Chrom im durchströmten Aquifer. Berichte, Fachbereich Geowissenschaften, Universität Bremen 101: 135 pp.

EDWARDS, E.A. & GRIBIC-GALIC, D. (1994): Anaerobic degradation of toluene and o-xylene by a methanogenic consortium. Appl. Environ. Micribiol. 60: 313-322.

EDWARDS, E.A.; WILLS, L.E.; REINHARD, M. & GRIBIC-GALIC, D. (1992): Anaerobic degradation of toluene and xylene by aquifer microorganisms under sulfate-reducing conditions. Appl. Environ. Microbiol. 58: 794-800.

GILLHAM, R.W. & O'HANNESIN, S.F. (1994): Enhanced degradation of halogenated aliphatics by zero-valent iron. Ground Water 32/6: 958-967.

HELLMOND , P. & USAKOW, D.F. (1992): Anorganische Korrosionsschutzschichten – Struktur und Eigenschaften. Deutscher Verlag für Grundstoffindustrie, Leipzig, 296 p.

HENCKE, J. (1998): Redoxreaktionen im Grundwasser: Etablierung und verlagerung von Reaktionsfronten und ihre bedeutung für die Spurenelement-Mobilität. – Berichte, Fachbereich Geowissenschaften, Universität Bremen 128: 122 p.

HERON, G.; CHRISTENSEN, T.H. & TJELL, J.C. (1994a): Oxidation capacity of aquifer sediments. Environ. Sci. Technol. 28: 153-158.

HERON, G.; CROUZET, C.; BOURG, A.C. & CHRISTENSEN, T.H. (1994b): Speciation of Fe(II) and Fe(III) in contaminated aquifer sediments using chemical extraction techniques. – Environ. Sci. Technol., 28: 1698-1705.

KÖLLING, M. (1986): Vergleich verschiedener Methoden zur Bestimmung des Redoxpotentials natürlicher Wässer. Meyniana 38: 1-9.

LINDBERG, R.D. & RUNNELLS, D.D. (1984): Ground water redox reactions: An analyses of equilibrium state applied to E_H measurements and geochemical modeling. Science 225: 925-927.

LONG, G.G.; KRUGER, J.; BLACK, D.R. & KURIYAMA, M. (1983): EXAFS study of the passive film on iron. Journal Electrochemical Society 130/1: 240-242.

LOVLEY, D.R. & LONERGAN, D.J. (1990): Anaerobic degradation of toluene, phenol and p-cresol by the dissimilatory iron-reducing organisms. Appl. Environ. Microbiol. 56: 1858-1864.

MATHESON, L.J. & TRATNYEK, P.G. (1994) Reductive dehalogenation of chlorinated methanes by iron metal. Environ. Sci. Technol. 28/12: 2045-2053.

O'HANNESIN, S.F. & GILLHAM, R.W. (1998): Long-term performance of an in-situ "iron wall" for remediation of VOCs. Ground Water 36/1: 164-170.

ODZIEMKOWSKI, M.S. & GILLHAM, R.W. (1997): Surface redox reactions on commercial grade granular iron (steel) and their influence on the reductive dechlorination of solvent. Micro raman spectroscopic studies. 213th American Chemical Society National Meeting, preprints of extended abstracts. 37/1: 177-180.

PARKHURST, D.L.; THORSTENSON, D.C. & PLUMMER, L.N. (1990): PHREEQE - A Computer Program for Geochemical Calculations. (Conversion and Upgrade of the Prime Version of PHREEQE to IBM PC-Compatible Systems by J.V. Tirisanni & P.D. Glynn). - U.S. Geol. Survey Water Resour. Invest. Rept. 80-96, Washington D.C.: 195 pp.

SCHLICKER, O.; EBERT, M.; SCHAD, H.; WÜST, W. & DAHMKE, A. (1998): Geochemische Modellierungen zur Abschätzung der Grundwassergüte im Abstrom von Fe^0-Reaktionswänden. Terra Tech 1: 43-46.

SCHUHMACHER, T. (1995): Identification of Precipitates formed on zero-valent Iron in anaerobic aqueous Solutions. unpublished MSc. Thesis, University of Waterloo.

SHUEY, R.T. (1975): In Development in Geology 4. Semiconducting ore minerals. 371-388.

STRATMANN, M. & MÜLLER, J. (1994): The mechanism of the oxygen reduction on rust-covered metal substrates. Corrosion Science 36/2: 327-359.

STRATMANN, M. & HOFFMANN, K. (1989): In-situ Mössbauer spectroscopic study of reactions within rust layers. Corrosion Science 29/11/12: 1329-1352.

STUMM, W. & MORGAN, J.J. (1996): Aquatic Chemistry: Chemical equilibria and rates in natural waters, 3rd ed., John Wiley & Sons Inc., New York: 1022 pp.

SWEENY, K.H. & FISCHER, J.R. (1972): Reductive degradation of halogenated pesticides. U.S. Patent No. 3: 640, 821.

VON GUNTEN, U. & ZOBRIST, J. (1993): Biogeochemical changes in groundwater-infiltration systems: Column studies. Geochim. Cosmochim. Acta 57: 3895-3906.

ZEYER, J.; KUHN, E.P. & SCHWARZENBACH, R.P. (1986): Rapid microbial mineralization of toluene and 1,3-Dimethylbenzene in the absence of molecular oxygen. Appl. Environ. Microbiol. 52: 944-947.

Chapter 18

Microbial Metabolism of Iron Species in Freshwater Lake Sediments

B. Schink & M. Benz

18.1 Introduction

Sediments develop by sedimentation of organic and inorganic residues of primary and secondary production as well as by inorganic precipitates, e.g., metal hydroxides, carbonates, silicates, and phosphates. The accumulation of this material at the bottom of freshwater lakes leads to an intensification of mainly microbial degradative activities which oxidise and transform the organic freight with concomitant reduction of oxygen and other electron acceptors. It is the activity of micro-organisms, especially of bacteria, which leads to the reduction of available electron acceptors, to an accumulation of reduced derivatives, and with that to changes of the redox potential in such sediments.

The basic processes involved in the degradation of organic matter by such microbial communities are known for a long time. As long as molecular oxygen is available it acts as the preferred electron acceptor, followed by nitrate, manganese(IV) oxide, iron(III) hydroxides, sulfate, and finally CO_2 with the release of nitrite, ammonia, dinitrogen, manganese(II) and iron(II) carbonates, sulfides, and finally methane as products of microbial reductive activities (STUMM & MORGAN,

1981). These preferences for the various acceptor systems are mainly determined by the redox potential and the availability of the redox systems under consideration, with the most positive ones at the beginning and the lower ones to the end, according to the scheme depicted in Table 18.1.

Tab. 18.1: Preferred redox potential ranges for the dominant microbial redox transformations in a freshwater lake sediment. After ZEHNDER & STUMM (1988) and numerous other sources.

Redox process	*Redox potential* [V]
Nitrate reduction	0.5 to 0.2
Manganese(IV) reduction	0.4 to 0.2
Iron(III) reduction	0.2 to 0.0
Sulfate reduction	0 to -0.15
CO_2 reduction	-0.15 to -0.22

Reduction of these electron acceptors with electrons from organic matter (average redox potential of glucose → 6 CO_2: -0.434 V; calculated using data from THAUER et al., 1977) provides metabolic energy in the mentioned sequence, and this sequence of preference is also translated via diffusive transport of the dissolved electron carrier systems into a spatial order in the sediment from the top to the bottom, with oxygen respiration at the sediment surface and methanogenesis in the deepest layers. The organic matter is to some extent oxidised to CO_2 but is also transformed into polymeric derivatives, similar to humic material in terrestrial ecosystems. These humic compounds (fulvic acids, "Gelbstoffe") are the dominant fraction of organic matter in lake water and persist in the sediment as the most important organic fraction. They contain aromatic and aliphatic residues from all chemical constituents of biomass, e.g. phenolic compounds, proteins, carbohydrates, fats, and even macromolecular cell structures as huge as entire murein sacculi. The longer this digestion process proceeds the less amenable this material will be to biochemical oxidation. Recent work in our lab has shown that proteinaceous humic constituents are degraded faster than others, shifting the total composition of the remnant material to a more aromatic character (KAPPLER et al., in prep.). Finally, hydroquinone as a model component for humic material releases electrons at an average redox potential of -0.328 V during its oxidation to CO_2 (glucose at -0.434 V, see above), indicating that conversion of such material to, e.g., methane and CO_2 ($E^{\circ\prime}$= -0.24 V), approaches not only kinetic but also energetic limitations.

The spatial sequence of redox processes in sediments on the basis of these considerations is preserved only in low-lying profundal sediments at water depths below 20 m; littoral sediments or sediments of shallow lakes are less stable and are subject to irregular mixing by wave action or bioturbation.

The redox potential measurable, e.g. by a platinum electrode, is determined mainly by reactive sediment constituents such as O_2, Fe^{2+} or Mn^{2+} ions, or hydrogen sulfide. Other carriers of low reactivity such as NH_4^+, iron hydroxides or carbonates, MnO_2, sulfate or methane do not directly influence the measurement of the redox potential. However, they do so indirectly in the presence of suitable catalysts, e.g. micro-organisms that are metabolically active.

The microbial activities involved in these redox processes are well-known as far as dissolved electron carrier systems are concerned, because organisms dealing with such carriers are easy to cultivate in the laboratory. Far less is known about the transformation of substantially insoluble electron acceptors such as manganese or iron compounds, due to their low solubility and difficulties in handling in the laboratory. In the present communication, we want to concentrate on new information on microbial activities involved in the transformation of iron compounds in lake sediments, and their possible impact on measuring the redox potential.

18.2
Iron Compounds in Lake Constance Sediments

Since iron(III) hydroxides and -oxihydroxides displaying extremely low solubility the free water column of oxygenated lake water contains iron mainly as constituents of living organisms or in complexes of biological origin, and after degradation of the organic residues iron precipitates as iron(III)hydroxides, -oxihydroxides, carbonates, silicates, or phosphate. These compounds accumulate in the sediments to high concentrations, averaging at around 1 to 5 % of the sediment dry matter (20 to 100 mM), depending on the type of lake and the chemistry of its catchment area. With these high concentrations, iron(III) is a very important electron acceptor in the sediment, compared to oxygen (around 0.3 mM), nitrate (<0.1 mM), or sulfate (around 0.2 mM in many freshwater lakes). Reduction of ferric iron hydroxides etc. leads to release of Fe^{2+} ions and the formation of ferrous carbonate (siderite), depending on the activity of bicarbonate and carbonate. The Fe^{2+} ion can diffuse through the sediment and may transfer electrons to other oxidised ferric iron or manganese minerals; they are also a redox-active species that contributes to the redox potential measured by the platinum electrode. However, their concentration is usually low and limited by the solubility of siderite. Besides the nano-particle movement, the substantially insoluble iron minerals are immobile and do not interfere with diffuse transport processes.

The distribution of Fe^{2+} in the pore water, and HCl-soluble (1 M) Fe(II) and Fe(III) in a profundal sediment core from Lake Constance is shown in Figure 18.1a, in comparison to an oxygen profile in Figure 18.1b. Oxygen profiles were measured with micro-electrodes; iron was quantified with the ferrozine method in extracts of slices of the sediment core obtained under a dinitrogen gas atmosphere before and after HCl (1 M) extraction and reduction with hydroxylamine. Oxygen is depleted by microbial reduction in the upper 5 cm of the sediment, with maximal depletion activity (highest slope) at 4 to 5 cm depth. Dissolved Fe^{2+} ions reach a concentration maximum (0.04 mM) at 4 to 7 cm depth, decreasing upwards as a result of chemical or microbial reoxidation, and down-

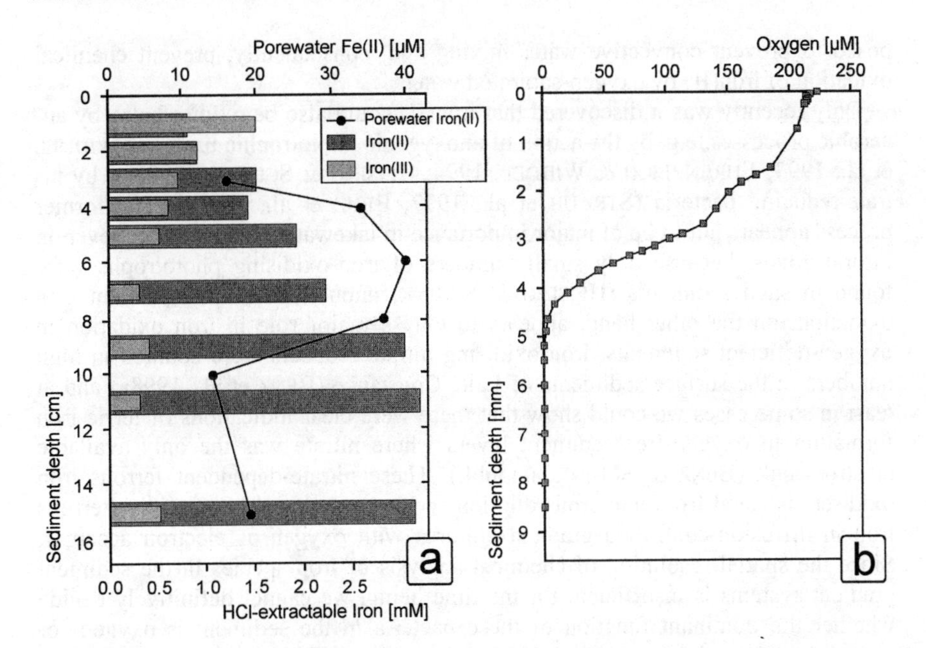

Fig. 18.1: Profiles of distribution patterns of iron species (a) and oxygen (b) in a sediment core taken July 17, 1996 in Lake Constance, Überlinger See, off the shore of Wallhausen, at 130 m water depth.

wards probably due to precipitation. In the non-water-soluble, HCl-extractable fractions, iron(III) dominated in the upper sediment layers whereas the reduced forms became dominant at depths lower than 4 cm. Nonetheless, there was always a small fraction of oxidised iron present even in lower sediment layers, indicating that this iron(III) was not accessible to microbial reduction. Comparison with similar profiles from more acidic fresh water lake sediments could provide information on whether this limited efficiency of iron reduction is due to competition against siderite formation. The increase of the total HCl-extractable iron content with depth from 2.0 to 3.5 mg/g may be a consequence of calcite dissolution and decomposition of the organic matter.

18.3 Microbial Oxidation of Iron Compounds

Aerobic bacterial oxidation of iron(II) is a well-known phenomenon and is catalysed in acidic environments by *Thiobacillus*-like bacteria, at neutral pH by morphologically conspicuous, filamentous bacterial forms such as *Gallionella* or *Leptothrix* species (SCHLEGEL, 1994). The latter form fluffy or slimy structures in springs and surface waters with high input of Fe^{2+}, and these structures are supposed to prevent convective water mixing and, consequently, prevent chemical oxidation of iron(II) by oxygen-saturated water.

Only recently was it discovered that iron(II) could also be oxidised also by anaerobic processes, e.g. by the action of anoxygenic phototrophic bacteria (WIDDEL et al., 1993; EHRENREICH & WIDDEL, 1994; HEISING & SCHINK, 1998) or by nitrate-reducing bacteria (STRAUB et al., 1997; BENZ et al., 1998a). The former process appears not to be of major importance in lakewater sediments, not even in littoral zones, because only small numbers of iron-oxidising phototrophs were found in such sediments (HEISING & SCHINK, unpubl.). Nitrate-dependent iron oxidation, on the other hand, appears to play a major role in iron oxidation in oxygen-deficient sediments: iron-oxidising nitrate reducers were counted in high numbers in the surface sediments of Lake Constance (BENZ et al., 1998a) and at least in some cases we could show that there were clear indications of ferric iron formation in oxygen-free sediment layers where nitrate was the only available electron sink (BENZ & SCHINK, unpubl.). These nitrate-dependent ferrous iron oxidisers isolated from numerous dilution series were also able to oxidise ferrous iron in diffusion-controlled gradient cultures with oxygen as electron acceptor. Since the spatial resolution of chemical analysis of iron species in the sediment gradient systems is insufficient for the time being, we cannot definitively decide whether the dominant function of these bacteria in the sediment is oxygen- or nitrate-dependent iron oxidation. Nonetheless, the profiles given in Figure 18.1a and b clearly indicate that iron(III) dominates to a large extent in the upper sediment layers that are entirely depleted of oxygen.

These bacteria were inconspicuous in shape, simple, short and rod-shaped, not forming any kind of unusual hypercellular structures. Since the sediment provides a structured environment impeding convective mixing of water from oxygen-containing and from lower layers rich in reduced iron, such aerobically iron-oxidising bacteria do not need to structure their environment themselves if they inhabit the sediment at the transition zone between oxygen-supplied and reduced layers. Perhaps these recently discovered iron oxidisers are far more important for ferrous iron oxidation on a global scale than the well-known and well-described *Gallionella*, *Sphaerotilus*, and *Leptothrix* species.

18.4 Reduction of Ferric Iron Hydroxides

Reduction of ferric to ferrous iron is chemically easy; the difficulty in microbial iron reduction is the extremely low solubility of ferric iron hydroxides. Ferric iron cannot be taken up by bacteria as such; aerobic bacteria have developed very refined and specific complexing agents (siderophores) to secure their iron supply for assimilatory purposes (NEIDHARDT, 1989). Such extremely energy-consuming systems cannot be applied by bacteria which use ferric iron as an electron acceptor in their energy metabolism. Delivery of electrons to insoluble ferric iron minerals would require immediate attachment of the iron-reducing bacterial cell to the mineral surface, but electron transfer between two non-dissolved particles becomes difficult even over extremely short distances. It was suggested recently that humic compounds could act as electron carriers between iron-reducing bacteria and insoluble iron minerals (LOVLEY et al., 1996). This concept is worth to be examined in more detail because humic compounds are actually present in sediments at comparably high concentrations (up to 1 % of total dry matter) and redox reactions between humic compounds and ferric iron minerals have been described repeatedly in the literature (SZILAGYI, 1970). Moreover, we could show recently that several fermenting bacteria can also use humic acids as electron acceptors (BENZ et al., 1998b), and electron transfer through humic compounds to iron(III)-oxides may therefore be a rather widespread type of respiratory activity not confined to typical iron-reducing bacteria.

We demonstrated recently that the iron-reducing bacterium *Geobacter sulfurreducens* excretes during growth significant amounts of a soluble c-type cytochrome into the growth medium. This cytochrome has a standard redox potential at -0.167 V and reduces iron(III)-hydroxide at very high rate. Thus, it acted as an extracellular iron(III)-hydroxide reductase and transferred electrons by diffusion over limited distances (SEELIGER et al., 1998). The same cytochrome could also mediate electron transfer to manganese(IV)oxide and to partner bacteria. Calculations of diffusion kinetics, actual cytochrome concentrations in growing cultures and growth rates gave reasonable evidence that this cytochrome contributed to a significant part to the iron reduction activity of growing cultures in the laboratory, and should do so *in situ*, in sediments and hydromorphic soils as well. Although this has not yet been tested, it appears reasonable to argue that such extracellular cytochromes would also react with platinum electrodes and would therefore help to characterise redox potentials of microbial communities in sediments to such monitoring devices.

18.5 Conclusions

This contribution concentrated on the transformation of iron compounds by new, so far unknown microbial activities. By mass, iron minerals represent the most important electron carrier system in most freshwater sediments. Due to their low solubility they do not interfere directly with the redox potential measured with the platinum electrode but do so only indirectly through Fe^{2+} which is present in the Fe(III)/Fe(II) redox transition zone at maximal albeit low concentrations (in the micromolar range). Humic compounds and extracellular cytochromes of certain iron-reducing bacteria may contribute further to the transfer of electrons between iron minerals, iron-metabolising bacteria, and platinum electrodes as monitoring devices. Oxidation of ferrous iron in freshwater sediments may be catalysed to a large extent in the oxygen-free layer of sediments with nitrate as oxidant by nitrate-reducing bacteria which have been described only recently and are able to use also oxygen as electron acceptor.

18.6 Acknowledgements

The authors are indebted to Horst D. Schulz, Bremen, for several fruitful discussions on redox processes in sediments. Experimental work in our lab was financed by the Deutsche Forschungsgemeinschaft (DFG) as part of its Special Research Project (SFB) "Cycling of matter in Lake Constance".

18.7 References

BENZ, M.; BRUNE, A. & SCHINK, B. (1998a): Anaerobic and aerobic oxidation of ferrous iron at neutral pH by chemoheterotrophic nitrate-reducing bacteria. Arch. Microbiol. 169: 159-165.

BENZ, M.; SCHINK, B. & BRUNE, A. (1998b): Humic acid reduction by *Propionibacterium freudenreichii* and other fermenting bacteria. Appl. Environ. Microbiol. 49: 4507-4512.

EHRENREICH, A. & WIDDEL, F. (1994): Anaerobic oxidation of ferrous iron by purple bacteria, a new type of phototrophic metabolism. Appl. Environ. Microbiol. 60: 4517-4526.

HEISING,S. & SCHINK, B. (1998): Phototrophic oxidation of ferrous iron by *a Rhodomicrobium vannielii* strain. Microbiology 144: 2263-2269.

LOVLEY, D.R.; COATES, J.D.; BLUNT-HARRIS, E.L.; PHILIPPS, E.J.P. & WOODWARD, J.C. (1996): Humic substances as electron acceptors for microbial respiration. Nature 382: 445-448.

NEILANDS, J.B. (1989): Siderophore systems of bacteria and fungi. In: BEVERIDGE, T.J. & DOYLE, R.J. (Eds): Metals ions and bacteria, Wiley and Sons, New York, pp 141-163.

SCHLEGEL, H.G. (1994): Allgemeine Mikrobiologie. Thieme Stuttgart.

SEELIGER, S.; CORD-RUWISCH, R. & SCHINK, B. (1998): A periplasmic and extracellular *c*-type cytochrome of *Geobacter sulfurreducens* acts as ferric iron reductase and as electron carrier to other acceptors and to partner bacteria. J. Bacteriol. 180: 3686-3691.

STRAUB, K.L.; BENZ, M.; WIDDEL, F. & SCHINK, B. (1996): Anaerobic, nitrate-dependent microbial oxidation of ferrous iron. Appl. Environ. Microbiol. 62: 1458-1460.

STUMM, W. & MORGAN, J.J. (1981): Aquatic chemistry. John Wiley & Sons New York.

SZILAGYI, M. (1971): Reduction of Fe^{3+} ion by humic acid preparations. Soil Sci. 111: 233-235.

THAUER, R.K.; JUNGERMANN, K & DECKER, K. (1977): Energy conservation in chemotrophic anaerobic bacteria. Bacteriol Rev. 41: 100-180.

WIDDEL, F.; SCHNELL, S.; HEISING, S.; EHRENREICH, A.; AßMUS, B. & SCHINK, B. (1993): Anaerobic ferrous iron oxidation by anoxygenic phototrophs. Nature 362: 834-836.

ZEHNDER, A.J.B. & STUMM, W. (1988): Geochemistry and biogeochemistry of anaerobic habitats. In: ZEHNDER, A.J.B. (Ed.): Biology of anaerobic microorganisms, John Wiley and Sons New York, pp. 1-38.

Chapter 19

Redox Measurements in Marine Sediments

H.D. Schulz

19.1 The Scope of Redox Measurements

Almost all (bio)geochemical processes in a young marine sediment are directly, or at least indirectly, bound to the decomposition of organic matter. Organic matter is synthesised in the euphotic zone of the water column as far as the sunlight is able to pervade. Only a small part of the primary production is buried in the sediment (SUESS et al., 1985) where it, functioning as an energy carrier, constitutes the motor of early diagenetic processes (BERNER, 1980).

The decomposition of organic matter proceeds in a succession of redox reactions oxidising the organic substance to yield carbon dioxide and water. The first electron acceptor is free oxygen dissolved in pore water, followed by manganese present in the sedimentary solid phase, then dissolved nitrate, then iron oxides of the solid phase, and finally sulfate dissolved in the pore water fraction. In the end, a fermentation reaction occurs in which one half of the zero-valent carbon bound in the organic substance is converted into the (-4)-valent carbon of methane, while the other half is converted into the (+4)-valent carbon of carbon dioxide. FROELICH et al. (1979) published a reaction scheme that includes the amounts of nitrogen and phosphorus bound in the organic substance (Figure 19.1). The

C:N:P-ratio of the organic substance was assumed to be 106:16:1, thus corresponding to the average ratio of these elements usually found in marine organic substances (*Redfield-Ratio*, after REDFIELD, 1958).

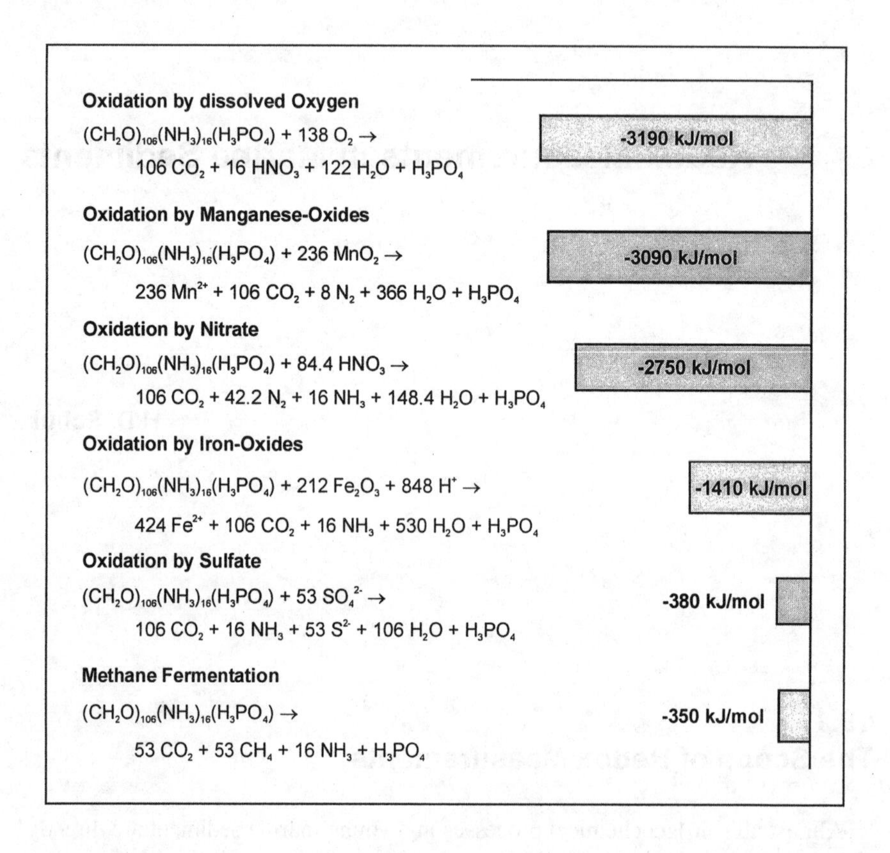

Fig. 19.1: Schematic representation of redox reactions in the biogeochemical decomposition of organic substances in marine sediments, according to FROELICH et al. (1979). The reactions are shown in the order of decreasing energy yields for the micro-organisms involved.

Accordingly, these reactions account for a succession of zones formed in the young marine sediment. The redox environment of each zone is determined by the redox reactions involved in the decomposition of the buried organic substance:

- Into the zone nearest to the sediment surface, dissolved free oxygen is transported by diffusion from the mostly oxic bottom water, or it is transported into the sediment by the action of macro-organisms living therein. In this zone, the organic substance is predominantly oxidised by the available amount of free oxygen. In the course of these reactions, carbon dioxide, nitrate and phosphate are released into the pore water.

- In the zone immediately below, manganese(IV)oxides serve as electron acceptors in the decomposition of organic substance. Once it becomes released into the pore water, divalent manganese is mainly carried upwards by diffusion, bioirrigation, or bioturbation into the zone characterised by the presence of free oxygen, where it is re-oxidised to manganese(IV)oxide. Sedimentary manganese consequently belongs to a cyclic movement, determined by various redox reactions, which altogether contribute to the downward transport of oxygen into deeper layers of the sediment.
- In the subsequent zone, nitrate which had been previously released in the oxygen zone serves as the main electron acceptor in the oxidation of organic substance. Most frequently, nitrate is reduced to elementary nitrogen (N_2-gas). This process is not very essential in quantitative respect, and it only converts oxygen which, ultimately, also originates from the uppermost oxygen zone.
- Of quantitative importance is rather a group of reactions in which various oxides of trivalent iron serve to oxidise organic matter. The often high content of iron oxides in sediments bestows upon this group of reactions an importance similar to that of the reactions with free oxygen.
- Below, there is a zone in which sulfate, highly concentrated in marine pore water, functions as the electron acceptor. Recent publications (e.g. NIEWÖHNER et al., 1998) have demonstrated that this deep sulfate reduction is not accomplished in most sediments by organic matter, but by methane which diffuses upwards from still deeper sedimentary layers. The redox turnover rate is almost as high as in the oxygen zone.
- As yet, the processes of methane fermentation taking place in the deeper regions of marine sediments have not been sufficiently investigated regarding their biogeochemical reactions, wherefore no further comment will be made as to the lowest reaction zone of Figure 19.1.

All zones are characterised by the mentioned redox reactions and determined as to their redox environment. The redox environments range from +500 mV in the oxic zone down to about –200 mV in the sulfidic zone and methane zone. The dimensions of these zones essentially vary depending on the concentration of decomposable organic substance in the sediment, the delivery of electron acceptors and thus, ultimately, the turnover rates in the sediments. The measurement of a redox potential depth profile therefore provides information about these redox reactions in the young sediment and, on account of the dimensions of the specific redox zones, about the various turnover rates of the early diagenetic processes. The redox potential is therefore an important parameter in the marine sediment characterising its redox environment. The measurement of the redox potential is done as soon as possible after sampling and should always be a desirable parameter when the pore water/sediment system is studied.

19.2 How Measurements are Performed

Nowadays, the pH-value and the concentrations of dissolved oxygen, carbon dioxide and calcium can be measured *in-situ* with the aid of fine needle-type electrodes. Furthermore, the redox potential can also be measured with plunger electrodes. An *in-situ* application of oxygen micro-electrodes in marine sediments has been described particularly by REVSBECH et al. (1980), JØRGENSEN AND REVSBECH (1985), and REVSBECH (1989). GLUD et al. (1994) have described the successful application of micro-electrodes for measuring pH and oxygen in deep-sea sediments. The penetration depth into which these electrodes usually reach is only some few centimetres, therefore less sensitive and less fragile optodes are now preferentially used for the *in-situ*-measurement of oxygen (KLIMANT et al., 1995).

In other all cases in which deeper sedimentary zones are studied, *ex-situ* measurements have to be performed with sediment samples, or rather, in the centrifugally or by squeezing extracted pore water. Sediment cores measuring up to 20 meters in length are recovered from the marine environment with various types of coring devices, all of which have in common that they can be lowered down to the bottom of the ocean from the decks of research vessels by steel cables. These corers are applicable in any depth where they extract a relatively unperturbed sediment core of 10 to 20 cm in diameter (e.g. multi-corer, box-corer, gravity corer, piston-corer). Most frequently, this core will be enclosed in a plastic liner and later cut into several pieces measuring one meter in length, once it is retained on board. Both ends of these segments are then sealed with plastic caps. Further processing and storage of deep-sea core segments is carried out at *in-situ* temperature (between 2 and 4 °C).

Further core processing toward the preparation and analysis of pore water ensues once the sediment core has been recovered. Since not all segments of a particularly long core can be processed at once, some sealed core segments have to remain stored under *in-situ* conditions. Experience shows that this is possible for the duration of several days without risking any essential alterations of the core samples.

Opening of the sediment core is done in a glove-box, in an inert atmosphere (preferentially argon, but highly purified nitrogen does as well). The core is then sliced in halves, in order to permit the withdrawal of unperturbed material from the fresh cut surface. The redox potential and the pH-value are measured in the fresh sediment with plunger electrodes, immediately after the core is opened. Upon studying anoxic parts of the sediment, it has been demonstrated that even the slightest oxygen impurities within the inert atmosphere of the glove-box can affect the measurement of the redox potential considerably. Subsequently, samples are taken from which the pore water is extracted prior to analysis, either by squeezing or centrifugation.

In contrast to pH-electrodes, E_H-electrodes cannot be calibrated, or rather, the electrodes will „forget and forgive“ the immersion into fixed calibration solutions of standard potential only after several days. Whether an electrode works properly can be only ascertained when several tried and tested, well maintained and recently polished electrodes are used in parallel, allowing the comparison of the recorded values whenever there are cases of doubt. Some „residual uncertainty" will always remain attached to the measured values, however, the self-critical geochemist should already be familiar with this kind of uncertainty from other parameters.

19.3 Typical E_H-Profiles and their Interpretation

The following examples are taken from an extensive set of pore-water data, compiled in the special research project 261 „The South Atlantic in the Late Quaternary -Reconstruction of Material Balances and Current Systems", more exactly, the partial project A2 concerned with „Transformation Processes and Material Fluxes in Sediment/ Pore-Water systems". These were obtained in the time between 1988 and 1998 and are the result of several expeditions with the research vessel „Meteor" to the Southern Atlantic. Apart from publications already mentioned, this data set is also available in the internet at "http://www.pangea.de".

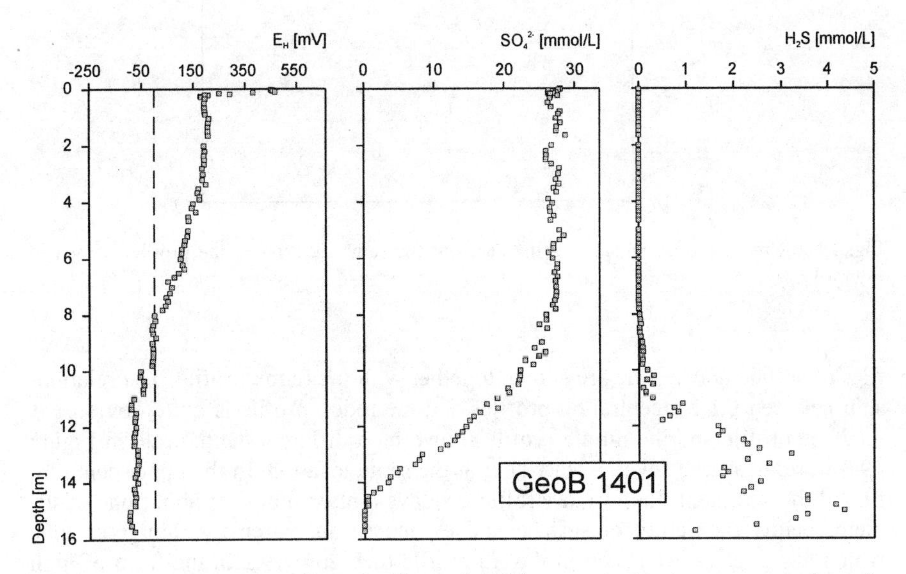

Fig. 19.2: E_H-profile with profiles of sulfate and sulfide concentration in pore water of a sediment core recovered from a water depth of approximately 4000 m, off the estuary of the Congo River at 6°56'S 9°00'E.

Figure 19.2 shows a redox profile of a sediment core which has been recovered as described from a water depth of approximately 4000 m, off the estuary of the Congo River (SCHULZ et al., 1994; SCHULZ & ZABEL, 1999). Concentration pro-

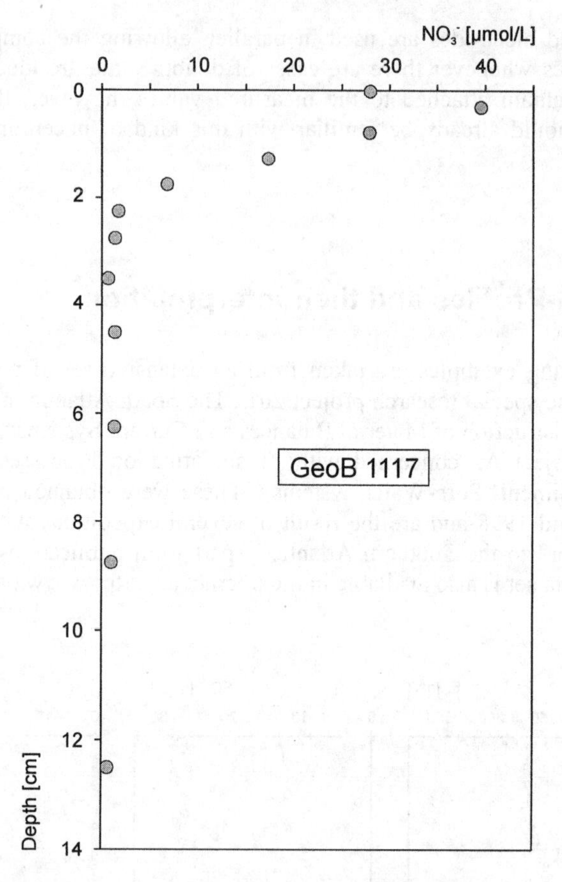

Fig. 19.3: Nitrate profile which was measured at the same location as the profiles shown in Figure 19.2.

files of sulfate and sulfide are shown together with the redox profile. The relationship between the concentration profiles and the redox profile is quite obvious. A methane profile and the nitrate profile shown on a different depth scale in Figure 19.3 were measured several years later at the same location. In the upper centimetres of the sediment, the nitrate profile displays a maximum just about one centimetre below the sediment surface and evidences an extremely thin oxic zone which has already been seen in the E_H-profile of Figure 19.2 in the form of high redox potentials in the uppermost range of the profile. Such a depth resolution for nitrate has not been possible during the earlier sediment sampling.

Distinctive redox potential zones are distinguishable in Figure 19.2. Looking from bottom to top, these are the following:

- In the depth range between 15 m and approximately 10 m below the sediment surface, the sulfidic zone is characterised by distinctly negative values of about –100 mV.

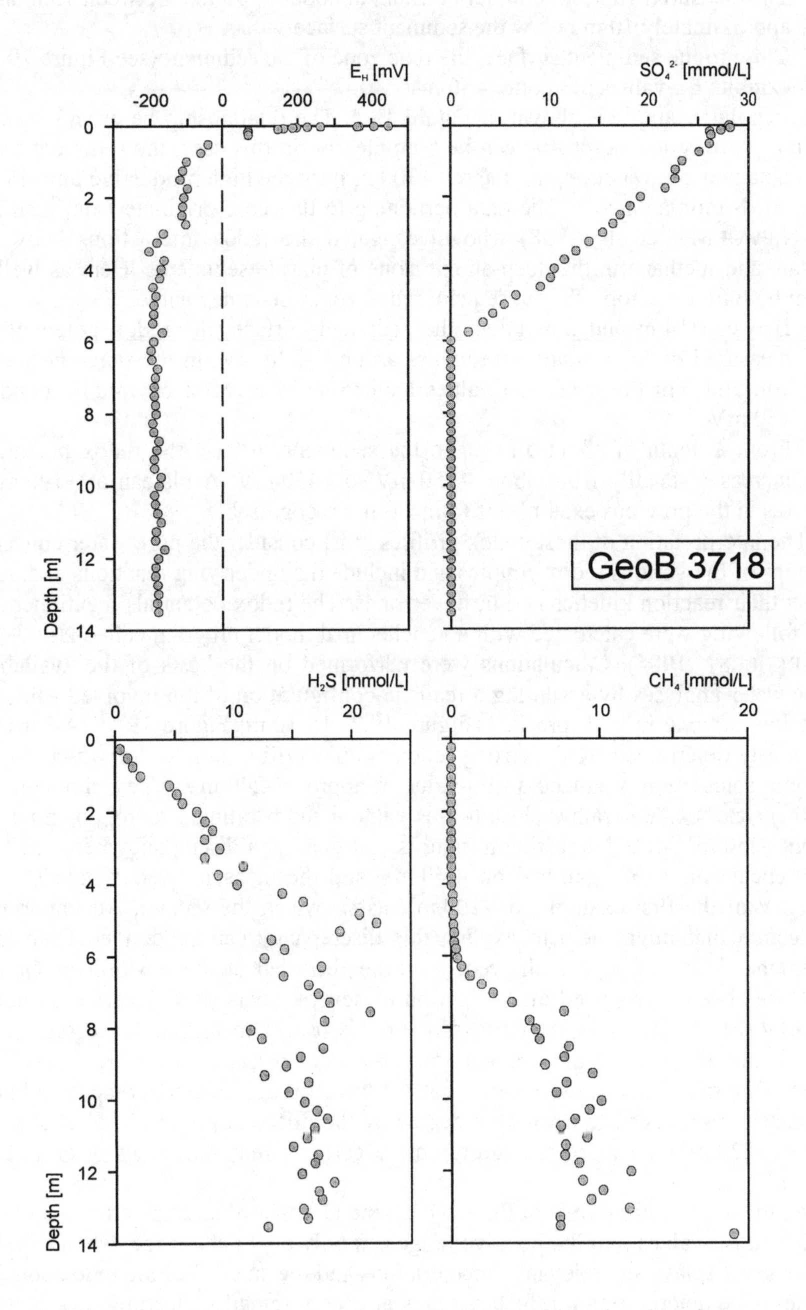

Fig. 19.4: E$_H$-profile with concentration profiles of sulfate, sulfide and methane in pore water of a sediment core recovered from a water depth of approx. 1300 m from an upwelling area off the coast of Angola at 24°54'S 13°10'E (modified from NIEWÖHNER et al., 1998).

- Between 10 m and approximately 3 m below the sediment surface there is a range in which the measured values rise continuously from about -100 mV to about +190 mV.
- The measured values are rather constant at about +190 mV between 15 m and approximately 10 m below the sediment surface.
- Close to the sediment surface, the oxic zone of the sediment (see Figure 19.3) exhibits E_H-values of about +450 mV.

A similar example is shown in Figure 19.4. The relationship between concentration profiles and E_H-profile can be seen clearly. In this case, the sediment core was obtained at a water depth of about 1300 m from the high productive upwelling area off Namibia/Angola. The data pertaining to this core originate from a study by NIEWÖHNER et al. (1998) who investigated the redox interactions between sulfate and methane in the deep-anoxic zone of marine sediments. Here as well - from bottom to the top - the subsequent redox zones become visible:

- Between 14 m and 5 m under the sediment surface, the redox potential is measured quite constant with values around –200 mV. In the range between 7 m and 5 m the measured values tend to show a redox potential around -150 mV.
- From a depth of about 5 m up to the sediment surface, the redox potential increases steadily from about –150 mV to +450 mV. A plateau at +190 mV (as in the previous example) is (almost) not recognisable.

The interpretation of these redox profiles must consider the pore water concentrations of the various redox couples and include the underlying reactions and, not least, their reaction kinetics and turnover rates. The redox potentials mentioned in the following were calculated with a geochemical model program called PHREEQC (PARKHURST, 1985). Calculations were performed on the basis of the available pore water analyses by assuming a realistic configuration of the involved solutes. The lowest zone in both profiles (Figure 19.2: 15-10 m; Figure 19.4: 14-7 m) is obviously determined by the strong redox couple sulfide/sulfate, for which equilibrium calculation produced an E_H-value of approx. –240 mV. The redox couple methane/carbonate is rather close to this value demonstrating a redox potential of about –250 mV. It is interesting to note is that there is a discrepancy between the theoretical values of –240 mV or –250 mV and the measured redox values of -100 mV of the first example, or -200 mV as shown in the second. An important indication that might help to explain this discrepancy can be derived from the additional information that the redox profile recorded at the position of GeoB 1401 has been reproduced after a period of several years (new location number GeoB 4914). In principle, the curve of the redox profile measured the second time had a similar shape as the previous one, but demonstrated lowest E_H-values at about -220 mV. This shows clearly that the lowest redox potentials measured are obviously reproduced to a variable degree by the different probes. The latter values of -220 mV are therefore surely „more correct" than values lying around -100 mV.

As to the example shown in Figure 19.2, the fact is well understood that an elevation of E_H-values into the positive range can only begin at a stage, when sulfide no longer displays any relevant concentrations and the sulfate/sulfide redox couple ceases to be determining for the environment. Yet, it remains uncertain

1. what ultimately causes the rather steady rise to a new plateau at about +190 mV,
2. what determines the redox potential in the intermediate depth ranges, and
3. why the rise, evident in the example of Figure 19.4, to values lying far in the positive range occurs at locations where sulfide is still abundant.

A plateau around the values of +190 mV as shown in Figure 19.2 (of which there is only a very slight indication in Figure 19.4) is mostly found when the core possesses a longer part of transition between its oxic and the anoxic segments, which is referred to as suboxic (related to the position) or postoxic (related to the time-dependent succession of the development) in the literature. It is believed that iron oxides in the sedimentary solid phase are mainly responsible for the decomposition of organic matter in this zone. This is supported by the fact that an E_H-value between +150 mV and +250 mV is calculated when, assuming various realistic conditions, there is a marked dominance of Fe^{2+} in the pore water solution and an equilibrium, for example, with regard to hematite. However, the individual reactions that determine the turnover rates and the redox milieu remain unknown. What is known about the complexity of the processes in the context of the early diagenesis of iron species in the marine sediment has been described by HAESE et al. (1997).

The oxic zone with E_H-values of about +450 mV is obvious in all profiles. As far as I know, it is agreed that the calculated equilibrium in the presence of free oxygen never reaches a value of +800 mV, but water having contact with the atmosphere actually always displays a redox potential of about +450 mV. In this regard, it may be important to know that if nitrate is present, an E_H-value of +350 mV will be calculated for the redox couple nitrate/ammonium, when realistic conditions are assumed. However, as the nitrate zone and the oxic zone (in a narrower sense exclusively referring to free oxygen) are, to some extent, quite identical, the measured potential may also frequently reflect a mixed potential at the electrode which is produced by both species combined.

19.4 Precision and Reproducibility

The question of precision and reproducibility of redox measurements directly conducted in the marine sediment has been marginally touched in the previous section. It might occur that redox electrodes in the sulfidic/anoxic zone will report values that are too high by 100 or 150 mV. Perhaps this erratic response can be avoided if the electrodes were given sufficient time for equilibration (hours or days?). Normally, one would abstain from such measure, because otherwise the sample to be measured would soon cease to be sufficiently anoxic, since, even in the argon atmosphere of the glove-box, the H_2S gas will be lost after the prolonged waiting periods. Furthermore, it may be that the value of about +450 mV persistently found in the oxic zone does not represent the equilibrium with free oxygen, instead, some other equilibrium or a mixed potential could be responsible for the effect.

Although most mV-measurement appliances even indicate one decimal, every one should be aware, at the latest upon waiting for a stable reading, that the „ones" and „tens" of the mV-display lie in the range of uncertainty. KÖLLING (1986) has described that a redox potential should only be accepted with an error margin of ±50 mV, even when conditions are optimal. Similar variances are found upon using plunger electrodes in marine sediments, especially when different electrodes are alternately applied in the measurements.

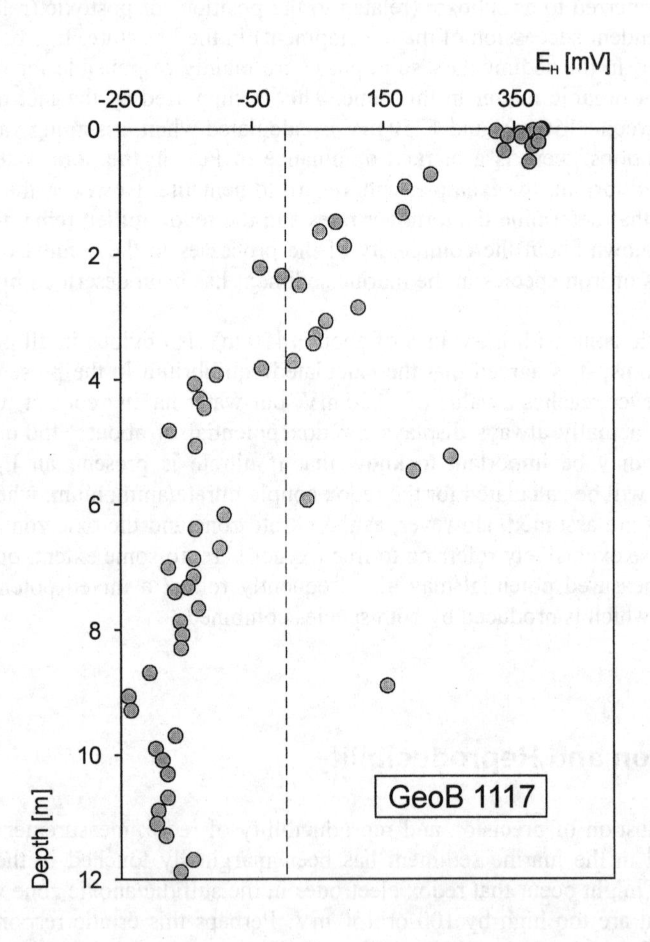

Fig. 19.5: E_H-profile of a sediment core recovered from an area of equatorial upwelling at a water depth of about 4000 m at 3°49'S 14°54'W (modified from SCHULZ et al., 1994).

Figure 19.5 shows a redox profile of a sediment core obtained from an area of equatorial upwelling in the South Atlantic. From this profile it can be clearly seen that the measured values scatter over the whole depth range at about ±50 mV. Nevertheless, this profile does make sense and its shape is comparable to other profiles in the concentrations from oxic zones (having values between +300 mV and +400 mV) to the anoxic/sulfidic zones (values down to -240 mV). Here as well, these measured values match well with the profile of sulfide (not shown) which indicated the presence of sulfide in the pore water fraction below about 6 m under the sediment surface.

According to the aforesaid, one might be inclined to consider the single positive values at about +200 mV, recorded at a depth of 5.5 m and again at 9 m beneath the sediment surface, as falsely measured. However, such outliers are indeed regularly found, especially in sediments that possess a heterogeneous pattern in the centimetre range. Such „spots“ containing a variable content of organic substance and/or iron were created at a time when the sediment layer in question had still been situated near the surface of the sediment and influenced by boring and grubbing fauna. Later, these trenches were filled with diverse material, experienced various diagenetic reactions, and hence produced different redox micro-environments.

One should not expect too much from any single redox measurement, yet, it would also be wrong to disqualify each "outlier“ as an outcome of a faulty measurement. The entirety of outlying values at a depth of 5m to 6m, as shown in the profile of Figure 19.5, the value measured at -200 mV as well as the comparison with other sediments, suggest that the redox potential reproduces a sedimentary redox milieu with a resolution which is higher than could ever be achieved by the analysis of a greater volume of extracted pore water.

19.5 Relevance of the Results

The unperturbed cores extracted from the marine sediment under the exclusion of atmospheric contact are only opened after they are secured in the argon atmosphere in the glove-box. Immediately after opening of the core, the measurement of the redox potential is carried out with the aid of freshly polished plunger electrodes made of platinum. For reason of comparison, several electrodes are employed in parallel.

When properly measured, the redox potential describes a milieu of the marine sediment which is determined by various (bio)geochemical processes. These milieu conditions extend from the oxic range near the surface of the sediment (values at about +350 mV to +450 mV) down to the deep anoxic/sulfidic range (values at about -200 mV to -250 mV). An intermediate suboxic or postoxic range (values at about +100 mV to +200 mV) is also frequently noticed.

From the measured redox potentials, it is surely neither adequate nor is it possible to derive any deliberate reaction taking place in the sediment or to draw conclusions on simple equilibria immediately present at the site of the electrode. Still, geochemical milieus can be discovered and characterised on the basis of an easy and quickly performed measurement. In individual cases, the redox potential may even record and reproduce densely clustered inhomogeneities much better than the normal analysis of squeezed-out pore water.

The redox potential is not a parameter which is precise down to a few percentage points of the measured value. Rather, frequently unexplainable fluctuations in the range of ±50 mV must be assumed. Yet, there are many other parameters which are even more difficult to measure in the pore water of marine sediments, with less precision and with less relevance (e.g. the dissolved gases H_2S, CO_2, CH_4), parameters no one would ever think of questioning.

19.6 References

BERNER, R.A. (1980): Early diagenesis: A theoretical approach.- Princeton University Press, Princeton, N.J., 241pp.

FROELICH, P.N.; KLINKHAMMER, G.P.; BENDER, M.L.; LUEDTKE, N.A.; HEATH, G.R.; CULLEN, D.; DAUPHIN, P.; HAMMOND, D.; HARTMAN, B. & MAYNARD, V. (1979): Early oxidation of organic matter in pelagic sediments of the eastern equatorial Atlantic: Suboxic diagenesis. Geochim. Cosmochim. Acta, 43: 1075-1090.

GLUD, R.N.; GUNDERSEN, J.K.; JØRGENSEN, B.B.; REVSBECH, N.P. & SCHULZ, H.D. (1994): Diffusive and total oxygen uptake of deep sea sediment in the eastern South Atlantic Ocean: In situ and laboratory measurements. Deep-Sea Res. 41: 1767-1788.

HAESE, R.R.; WALLMANN, K.; DAHMKE, A.; KRETZMANN, U.; MÜLLER, P.J. & SCHULZ, H.D. (1997): Iron species determination to investigate early diagenetic reactivity in marine sediments. Geochim. Cosmochim. Acta, 61 (1), 63-72.

JØRGENSEN, B.B. & REVSBECH, N.P. (1985): Diffusive boundary layers and the oxygen uptake of sediments and detritus. Limnol. Oceanogr. 30: 111-122.

KLIMANT, I.; HOLST, G. & KÜHL, M. (1995): Fiber-optic oxygen microsensors, a new tool in aquatic biology. Limnol. Oceanog. 40: 1159-1165.

KÖLLING, M. (1986): Vergleich verschiedener Methoden zur Bestimmung des Redox-Potentials natürlicher Wässer. Meyniana 38: 1-19.

NIEWÖHNER, C.; HENSEN, C.; KASTEN, S.; ZABEL, M. & SCHULZ, H.D. (1998): Deep sulfate reduction completely mediated by anaerobic methane oxidation in sediments of the upwelling area off Namibia. Geochim. Cosmochim. Acta 62: 455-464.

PARKHURST, D.L. (1995): User's guide to PHREEQC: a computer model for speciation, reaction-path, advective-transport, and inverse geochemical calculations. US Geol. Surv., Water-Resources Investigations Report 95-4227, 143pp.

REDFIELD, A.C. (1958): The biological control of chemical factors in the environment. Am. Sci. 46: 205-211.

REVSBECH, N.P.; JØRGENSEN, B.B. & BLACKBURN, T.H. (1980): Oxygen in the sea bottom measured with a microelectrode. Science 207: 1355-1356.

REVSBECH, N.P. & JØRGENSEN, B.B. (1986): Microelectrodes: their use in microbial ecology. In: MARSHALL, K.C. (ed.): Advances in Microbial Ecology, Vol. 9., Plenum, New York, 293-352.

REVSBECH, N.P. (1989): An oxygen electrode with a guard cathode. Limnol. Oceanogr. 34: 474-478.

SCHULZ, H.D. & ZABEL, M. (Eds.) (1999): Marine Geochemistry. Springer Verlag Berlin Heidelberg New York, 453pp.

SUESS, E. (1980): Particulate organic carbon flux in the ocean-surface productivity and oxygen utilization. Nature 288: 260-263.

Subject Index

Druck: Strauss Offsetdruck, Mörlenbach
Verarbeitung: Schäffer, Grünstadt